Off-Grid Solar Power Simplified

~

RVs, Vans, Cabins, Boats, and Tiny Homes

(Updated November 2024)

Introduction

To get access to color pictures and schematics in this book, visit my website at https://cleversolarpower.com/offgridsolarbook This page is exclusively for people who bought the book. Use the password on the last page of this book.

This book focuses on the practical approach to designing and installing an off-grid solar power system.

This book will start by discussing the basics of an off-grid system. This will include understanding the electrical units, measuring electricity, formulas, and the difference between AC and DC.

Next, we will discuss the different tools you need to build your system. This chapter is followed by a list of equipment that you will need. This list will save you many trips to the hardware store. The chapter also provides more detail about different fuses and where to put them.

From there, we dive into some circuitry and load types. The following chapter is the most practical because we will calculate how to size your off-grid system.

After you know how to size your system, we go more in-depth on choosing types of wire and wire sizes. This is a crucial step that cannot be overlooked.

Then we discuss the batteries, solar panels, charge controllers, and the different kinds of inverters.

We will then discuss grounding, neutral ground bond, using a ground fault current interrupter, galvanic corrosion in boats, and adding a generator.

Then, I will show you a step-by-step guide on making your solar system.

Lastly, I will provide you with some inspiration, schematics of different designs, and an explanation of fuse and wire sizes.

We end the book by recommending a list of quality brands for you to choose from.

This book is geared towards beginners and intermediate users and can be heavy on formulas. Don't let these intimidate you. They are easy to implement if you follow the step-by-step guidelines described in this book. Let's get started with the basic electrical units.

Electrical Units

Energy can be described in many ways, but basically, energy is referred to as the capacity to develop a specific work. Energy is presented in many ways in nature, and one of them is electricity, which is described as the capacity to establish electrical work.

To understand how electricity works, some important concepts must be addressed.

First, you must understand that an electrical circuit can be described as the interconnection of electrical components where at least three basic elements will exist:

- A power source
- A conductor
- A load

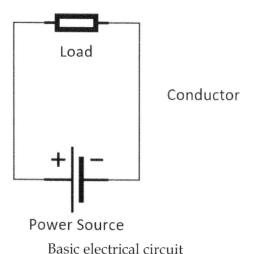

Basic electrical circuit

The power source is the element that produces or stores electricity (a battery, a generator, or a solar panel).
The conductor is the element through which electricity flows.

The load is the element that receives electricity for performing work (a lamp generates light, a motor provides motion, and an electrical resistor creates heat).

To understand the concepts used throughout this book, we need to describe some electrical terms first.

Voltage

Electricity is generated by the movement of electrical charges (electrons). In order to move an electron from one point to the other, it is necessary to perform electrical work.

This work is performed by an electromotive force (EMF) or voltage generated by the power source. Voltage can be understood as the pressure required to move the electrons from one point (A) to a second point (B) within an electrical circuit. The greater the voltage, the greater the flow of electrons through an electrical conductor.

This movement is generated from the highest electrical potential point (A) to the lower electrical potential point (B). Voltage is referred to as the electrical potential difference between these two points.

Voltage is measured in Volts (V).

Current

The second important electrical unit is current.

The electrical current can simply be understood as the intensity of the flow of electrons per second through a conductor.

This element is measured in Amps (A) or (I).

Resistance

The resistance refers to the opposition of a specific material to the flow of electrical current. In other words, the resistance provides a reference of how easy or how hard it is for the electrons to flow through any material (steel, aluminum, copper, etc.).

Every electrical load or conductor has an internal resistance, which is measured in ohms (Ω). For example, wood has a higher electrical resistance than copper.

Power

Power is one of the most important variables in electricity as it represents the combination of voltage and current in an electrical circuit. For an electrical load to perform any type of work (illumination, motion, heat), this element demands an instantaneous equivalent work source, which is provided by power. Power acts as a reference to the rate at which electricity is delivered (power source) or consumed (load) and is the product between voltage and current. The unit of power is Watt (W).

Watt-hour

As we mentioned before, power is the instantaneous rate at which electricity is provided or consumed. When we refer to energy, we are evaluating how electricity is being delivered or consumed over time.

In other words, electrical energy is described as the power generated or consumed over time.

As a general convention, electrical energy is expressed in watt-hour (Wh). Representing the consumption of a specific number of watts in a single hour. This unit will generally be used to account for the energy consumption of electrical loads in an off-grid solar power system.

Also, when consumption is higher, it is generally expressed in kilowatt-hours (kWh), which is the consumption of 1,000 watts of power in a single hour. You will be already familiar with the term Kilo Watt-hours, as it is listed on your electricity bill.

Amp-hour

Energy can also be expressed as the consumption of the amount of current in a single hour and is referred to as Amp-hour (Ah). Amp-hour is used to describe the amount of energy a battery can store at the nominal voltage. The battery of a typical smartphone has a capacity of 3 Amp-hours or 3,000 milli Amp-hours.

It's helpful to know that Ah doesn't say anything about the capacity of a battery. You need to know the voltage to know the total amount of watt-hours, as we will see later in the book.

Energy Measuring Equipment

Digital Meter

The digital meter (also known as a digital multimeter) is a test tool that is used to measure at least three variables:

- Voltage (AC and DC)
- Electrical current (DC)
- Resistance

A digital meter combines the capabilities of three tools into one: an analog voltmeter (measures volts), an analog ammeter (measures amps), and an analog ohmmeter (measures resistance).

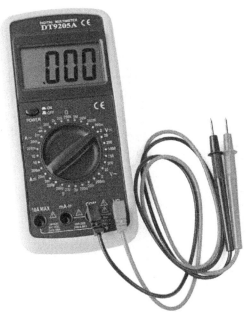

Typical digital meter

Every digital meter should have:

- A display to show measured values.
- A rotary switch to change variables.
- Input jacks for test leads.

The meter will have a range of unit scale measurements from millivolts (mV) to volts (V), from milliamps (mA) to Amps (A), and from milliohms(mΩ) to mega-ohms (MΩ).

You will need to know the unit's range that you will be testing to select the correct scale and obtain an accurate result.

Generally, we will use voltage, ohms, and sometimes amps for solar power applications.

Keep in mind that you will have test leads with insulated wires to test the electrical circuits. There will be a test lead for positive (red) and a test lead for negative (black). When measuring DC circuits, the colors (polarity) of the wires matter. It does not matter in AC because it is alternating. More on this later.

You need to put the black lead in the 'COM' input and the red lead in the 'V' input when you test voltage. Then you need to select the 'V' variable on the rotary switch and place the positive and negative leads accordingly to obtain an accurate measurement. Otherwise, you will get a negative value.

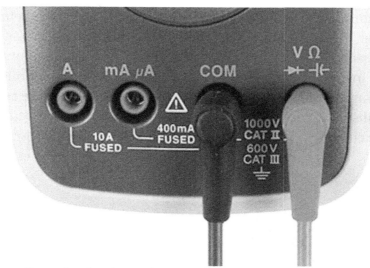

Input leads of a multimeter (left: black, right: red)

The same concept applies to measuring resistance. The black lead must go in the 'COM' input, and the red lead should be in the 'Ω' input, which is the same as the 'V' input. Select the Ω symbol on the rotary switch to measure the resistance of a device.

When testing voltage, you must measure it in an open circuit, which means measuring without load. For example, if you wish to measure the voltage that comes out of your power socket, you touch both pins' positive and negative to the electrical wires.

If you want to measure resistance, you must measure it without any applied voltage. For example, if you're going to check if a fuse is broken, you take out the fuse and measure at both ends of the fuse. If the display states a resistance of 1 or higher, the fuse is broken. Resistance is measured in an open circuit.

Although you could measure current using a digital meter, it's better to use an ammeter for this. This is because current will flow through your meter, potentially damaging or blowing a fuse inside if no load is applied. Most meters will be limited to 10 amps, which is not a lot. There is no need to measure current because you can calculate it using the formulas discussed in the next chapter.

1. Voltage AC (Volts)
2. Voltage DC (Volts)
3. Voltage AC (Milli Volts)
4. Resistance (Ohm)
5. Capacitance (uF)
6. Current (Amps)

The layout of a Fluke digital multimeter

Read the digital multimeter manual for further information on measuring voltage, resistance, and current.

Ammeter or Clamp Meter

The digital ammeter or clamp meter is a device that combines the advantages of a digital multimeter with an additional feature.

Like the digital multimeter, the ammeter can measure voltage (DC and AC), resistance, continuity, AC current, and other variables such as frequency, temperature, or capacitance.

The main difference with the multimeter is that the ammeter includes a clamp that allows you to measure the electrical current's RMS (root mean square) value. You need to open the clamps and close them around a conductor through which an electrical current is flowing. You cannot measure a cable with a positive and negative wire inside it. It can only measure one wire simultaneously because they will cancel each other out.

A cheaper ammeter can only measure AC current, not DC current. When you buy one, make sure it can measure DC current. Check the Amp rating to make sure it can read at least 100 amps (depending on your system).

Clamp meter by UNI-T

Basic Formulas

To perform calculations related to sizing an off-grid PV system, you need to use some basic formulas.

Ohms Law

$$V = I \times R$$

Where,

V = Voltage (Volts), sometimes written as 'U'
I = Electrical Current (Amps)
R = Resistance (Ohms)

Or,

$$I = \frac{V}{R}$$

Or,

$$R = \frac{V}{I}$$

Using a triangle for the formulas is an easy way to remember them. Using these formulas becomes easy once you remember the position of the units in the triangle.

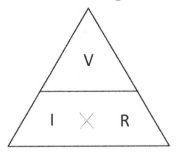

Power

$$P = V \times I$$

Where,

P = Power (Watts)
V = Voltage (Voltage), sometimes written as 'U'
I = Electrical Current (Amps)

Or,

$$V = \frac{P}{I}$$

Or,

$$I = \frac{P}{V}$$

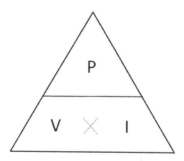

Energy or Watt-hours

$$E = P \times t$$

Where,

E = Energy (Watt-hours)
P = Power (Watts)
t = Time (hours)

Your electrical company uses this unit to bill your energy consumption. A watt-hour is a large number, so a kilowatt-hour is used. This means that 1,000 Watt-hours is equal to 1 kilowatt-hour or simply 1 kWh.

Running a heater with a power rating of 1,000 Watts for one hour will consume 1,000 Watt-hours of energy or 1 kWh.

$$E = 1,000\ Watts \times 1\ hour = 1,000\ Watt\ hours\ or\ 1\ kWh$$

Let's explore why voltage is an important factor when calculating Watt-hours.

Let's take two batteries for example.

- Battery one has a capacity of 2 Amp-hours at 1.2 Volts.
- Battery two has a capacity of 2 Amp-hours at 12 Volts.

These batteries seem to have the same amount of stored energy because the amp-hours are the same. However, this is not true because the voltage is different. Let's calculate the number of watt-hours stored in each of these batteries.

$$2\ Amp\ hours \times 1.2\ Volts = 2.4\ Watt\ hour$$

$$2\ Amp\ hours \times 12\ Volts = 24\ Watt\ hour$$

We can see that the stored energy in battery two is higher. That is why Amp-hours alone is not a clear indication of the available energy in a battery. Watt-hours is the correct unit.

Energy Cost

Every state or country has different electricity rates. To know how much you need to pay your electricity provider, you need to know your local electricity rate. The national average in the U.S. is $0.12 per kilowatt-hour. Note that the unit is per kilowatt-hour, not per watt-hour.

$$Energy\ Cost = Energy\ in\ kWh \times Rate\ in\ cents\ per\ kWh$$

If you run a light that has a power rating of 20 Watts and run it for 8 hours each day for 30 days, this is how much you will need to pay your electrical provider:

$$Energy\ in\ kWh = \left(\frac{20W}{1,000}\right) \times (8\ hours \times 30\ days) = 4.8kWh$$

$$Energy\ Cost = 4.8kWh \times \$0.12 = \$0.576$$

Amp-hours

One Amp-hour is equivalent to one amp expended for one hour.

$$Amp\ hours = Amps \times hours$$

Amp-hours are not commonly used in standard electrical practice but indicate a battery's capacity.

For example, a simple AA battery has a capacity of 2,000mAh (milliamp hours) or 2Ah. This means the battery can theoretically supply a load of two amps for one hour, one amp for two hours, and so on.

If you have a bigger battery with 100Ah capacity, you can draw 100 Amps for one hour or 10 Amps for 10 hours.

This is only in theory because different batteries have different depths of discharge and charge/discharge rates. We will discuss this more in the battery chapter.

Volt-Ampere

Volt-Ampere, also called VA, is the unit for apparent power. Some inverters list the VA value instead of the power (watts) value. Volt-Ampere and Power in DC systems are the same.

$$1W\ DC\ = 1VA$$

Volt-ampere becomes different when we transfer DC into AC and connect an inductive or capacitive load.

$$1W\ AC \neq 1VA$$

Most of your loads will be inductive, such as AC pumps or fans. You have to calculate the true power draw using the power factor of the specific device you want to power.

Let's say you want to power a pump of 800 watts with a power factor of 0.8. Then we apply the formula:

$$VA = \frac{P}{PF\ (power\ factor)} = \frac{800W}{0.8} = 1000VA$$

We would need an inverter capable of handling a load of 1000VA.

We can also calculate it in reverse:

$$W = VA \times PF$$

$$1000VA \times 0{,}8 = 800\,Watts$$

You can see that the Volt-Ampere rating is always higher than the true power rating. If you buy an inverter and attach inductive loads to it, you must be wary. If manufacturers state the inverter power in VA, you must multiply this number by 0,8 to have a more accurate power rating.

AC and DC

Difference Between AC and DC

In a solar power system, you will most likely have two signals:

- DC – Direct current
- AC – Alternating current

Therefore, you must understand the differences between these signals to know which device you must use, where, and why.

Simply put, DC stands for Direct Current, and AC stands for Alternating Current. The direct current is an electrical signal that constantly flows in a single direction across the electrical circuit over time.

AC   DC

AC and DC symbols

The following image shows a DC voltage signal that is stable and always has a constant value. However, values in other cases may increase or decrease over time.

DC Power Signal

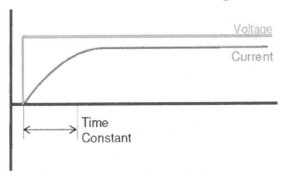

DC Voltage Signal
Source: LWClearning

The most important factor in qualifying as a DC signal is that the wave is either entirely positive or negative (always above or below the X-axis).

The best example of DC power source signals is a battery. Any battery provides a constant DC voltage (generally between 2V and 12V), while the electrical current may increase, decrease, or stay constant over time, depending on demand.

The solar panel is also a DC power source. When testing voltage and current for this component, you must use DC measurement instruments.

If you measure DC voltage with a digital meter, you will get a positive (+) or negative (-) reading. This way, you can figure out the polarity of a DC source. Try it with a battery and see if you can determine the polarities without looking at the positive and negative signs. Use your multimeter to double-check your solar panels' polarity before connecting them to your charge controller.

On the other hand, AC is entirely different. In this case, the electrical signal fluctuates. The signal changes between positive and negative values periodically over time, and this is what we call a sine wave.

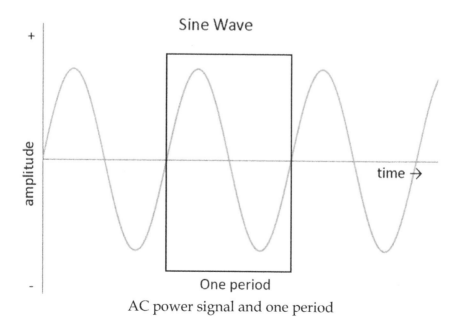

AC power signal and one period

This alternating shape is created by a synchronous generator, which uses mechanical energy (derived from the movement of a turbine driven by water or steam) and converts it into a periodic electrical signal output.

This generator has a rotor that spins around its axis 360 mechanical degrees. The output signal always features 360 electrical degrees. These changes in the value of the AC signal are referred to as the period of the wave, which is also referred to as the frequency of the signal.

Another method of creating an AC signal from a DC source is to use electronics, which we will discuss in the chapter 'inverters'. Most countries around the world use frequencies of 50 Hertz. The U.S., Canada, and several other countries use 60 Hertz, which means that electricity completes 60 periods in one second.

In the following image, you can see DC and AC voltage in the same graph. The wave is the AC signal, while the flat line is a DC signal.

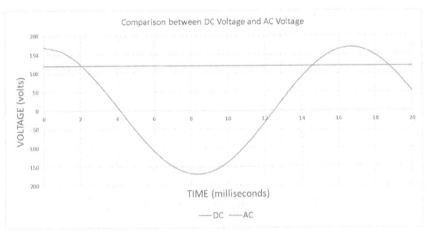

DC and AC Voltage Curves
Source: Zoroad Electric

You may now wonder how you can measure the value of the AC signal when it's constantly changing.

The measuring instruments for AC signals display a specific value called the Root Mean Square (RMS). If you measure with a digital meter, you will always measure the RMS unless specified otherwise.

The RMS is a constant value of the AC signal equal to the value of the direct current that would produce the same average dissipated power in a resistive load.

In the following figure, you can see the position of the RMS in an AC voltage signal.

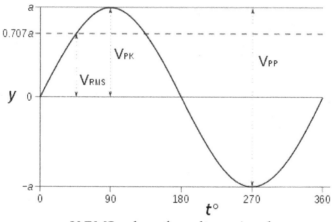

V RMS value of a voltage signal
Source: Wikipedia

The peak (VPK) of a sine wave in the U.S. measures 170 Volts AC, but you will see 120 Volts AC (VRMS) on your measuring device. Or,

$$Volts\ Peak = 1.414 \times Vrms$$

21

The power grid works using an AC signal that reaches the residential, commercial, and industrial sectors under particular specs related to quality, voltages, and ancillary services.

At the residential level, there may be three possible types of RMS values for the AC voltage:

- 120V single-phase
- 120/240V split phase
- 208V three-phase

Most countries use different power grids. In Europe, for example, the power grid is different. You will find these values:

- 230V single-phase
- 400V three-phase
- 690V three-phase

AC and DC in the System

A solar power system will most likely have both AC and DC signals. The solar panels, the batteries, and the charge controller will always work in DC. The inverter will transform the DC signal from the battery into an AC signal to power specific loads.

Therefore, when you test voltage or current in any part of the electrical circuit before the inverter, you must measure in DC. If you test any variable in a section after the inverter, you will measure AC.

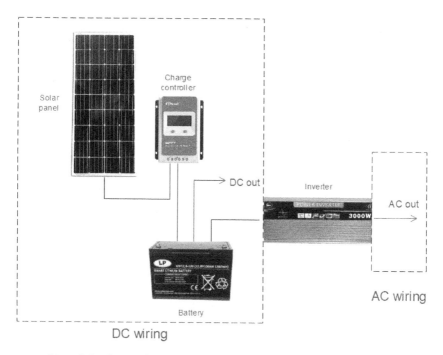

Solar panel

Charge controller

DC out

Inverter

AC out

Battery

AC wiring

DC wiring

Simplified graphic of DC and AC in the solar PV system

Tools

Wire Stripper

A wire stripper is a multi-use tool necessary for any electrical installation. It allows you to strip and cut any wire with gauges between 10-24AWG. The device will allow you to easily cut either copper or aluminum wires with precision without damaging the metal part of the electrical wire.

Wire stripper

Cable Stripper

A cable stripper is also needed to strip cables from #5AWG to 4/0AWG, which a wire stripper cannot do. The cable stripper can cut PVC, rubber, foamed polyethylene (PE), and other insulating materials.

This tool's interesting and useful feature is that it can make longitudinal, circular, spiral, and mid-span cuts to remove the cable's jacket. The device includes a cable holder that makes the cutting process easier.

You can adjust it quickly to the gauge of the cable. The cutting is made through a blade depth knob that adjusts the blade (which is also replaceable) to fit the size of the cable.

Always calibrate the tool on a wire end to ensure the blade doesn't cut the strands.

Cable stripper

Lug Crimping Tool

This product is intended for installations where battery banks are used. The product can crimp battery cable lugs with standard sizes between #8AWG and 1/0AWG.

The stripping of the cable to introduce the lug should be done with a cable stripper.

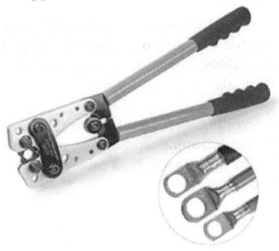

Lug crimping tool

Crimping lugs are crucial to your solar system. I suggest getting a hydraulic crimper if you need cables larger than 4AWG or 25mm².

You can also buy pre-crimped cables from currentconnected if you don't want to crimp them yourself.

Another option for the same purpose is a hammer lug crimper. However, I don't recommend using this tool because the crimping is not done optimally.

Hammer lug crimping tool

Wire Crimping Tool

This tool is suitable for crimping individual wires. It integrates a ratcheting mechanism, which is helpful for precise and repeatable crimps.

Its ratcheting mechanism lets you secure a wire connector before inserting the stripped wire into the lug. You can crimp wire terminals for gauge sizes between 22 and 10AWG (0.5-6mm²), split into three crimping options marked by red, blue, and yellow, indicating the gauge ranges for each purpose.

22 - 18 AWG I DIN 0.5 - 1.0 (mm²)

16 - 14 AWG I DIN 1.5 - 2.5 (mm²)

12 - 10 AWG I DIN 4.0 - 6.0 (mm²)

Wire crimping tool

Conduit Cutter

The following tool on our list is the conduit cutter. Conduit is generally used in electrical installations to protect cables or wires from water and physical damage.

However, to fit your wiring installation, you must cut the conduit to adjust the length properly. For this purpose, a conduit cutter tool is needed. The conduit cutter is used for multiple applications, such as cutting PVC or PEX pipe. It is also suitable for cutting CPVC, PP, and PE-XB pipes, allowing you to cut them within a few seconds.

PVC conduit cutter

If you have a vehicle, it is recommended that you use flexible polyethylene pipe. It will protect the wires from vibration and act as a cable highway, which is helpful if you want to add an extra wire from the vehicle's back to the front.

Screwdrivers

Screwdrivers are needed in almost any installation. However, an insulated screwdriver is preferred for electrical installations.

For this purpose, the best choice is to purchase a screwdriver set with six pieces tested to resist up to 1,000 Volts AC or 1,500 Volts DC. Each tool will be covered with a non-conductive material that can reach such a rating, making it safe for electrical installations.

Besides, a soft handle with an outer cushion grip adds 40% more torque than traditional plastic handles.

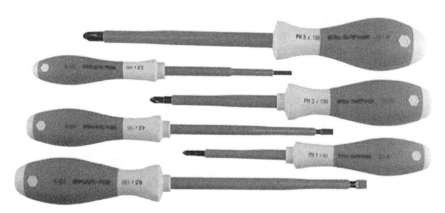

Electrical screwdrivers

Needle Nose Pliers

The needle nose pliers are the perfect tool to bend wires. Their half-round tapered jaws are longer and narrower, useful for places other pliers cannot reach.

These pliers come with a cutting tool but are generally not used for that purpose. The needle plier can usually be found in three handle styles: plastic-coated handles, comfort grip handles, and the 1000V insulated handle that meets IEC standards, which is the model for electrical installations.

Needle nose pliers

Wire Cutters

The wire cutter is another valuable tool for cutting individual wires. The device is integrated with machined jaws to provide maximum gripping strength and has an induction-hardened cutting edge that stays sharper for longer.

Normal wire cutter

Another valuable tool is a flush cutter. This tool will have a 21-degree angle for flush-cutting, and its heat-treated carbon steel construction can provide durability and long-life performance. The small size makes it easy to carry around.

Small wire cutter or flush cutter

Cable Cutter

There may be occasions when the wire cutter alone is not enough to perform all the necessary work for heavy-duty applications with thicker cables. The cable cutter is the perfect choice since it can cut up to 0AWG gauge cables (50mm²), both copper and aluminum.

Cable cutter

Hex Nut Ratchet Set

The hex nut ratchet set is something that you need to perform electrical installations of any kind. They can be used to tighten the battery terminals. They also come with a drive-bit set. The bits can be used to drive screws into the wall to mount components.

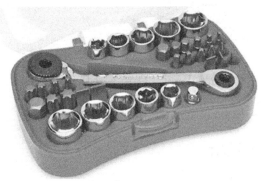

Ratchet set

Torpedo Level

Whenever you perform measurements to install devices or equipment, you need to keep the equipment straight. That´s when a level comes in handy.

Torpedo level

Hole Saw

The sharp teeth of each saw are perfect for making holes in wood, PVC boards, plastic, drywall, and even metal.

They can drill holes in your RV's roof to pass cables through.

Hole saw set

Cordless Drill

A cordless drill is suitable for drilling pilot holes and mounting screws or bolts on any wall surface. It is just another essential piece of your toolkit.

The cordless drill generally features two-speed transmission sets at low speed (about 500RPM) and high speed (about 1,900RPM), suitable for a wide range of drilling and fastening applications.

The most helpful feature is that this drill does not need an AC plug connection.

Cordless drill

Drill Bits

The drill bits are a valuable complement to the previous tool. They are needed to make pilot holes to mount appliances or devices on a wall surface, such as an inverter or a charge controller.

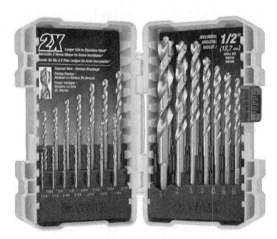

Drill bit set

Safety Goggles

Last but not least, you will need to wear safety goggles to protect your eyes during the installation of solar equipment.

You can generally find them in different sizes to fit the shape of your head. They can also integrate bi-focal shatterproof polycarbonate lenses. The lenses can be found in various colors, from copper to yellow, gray, smoke, and transparent. These lenses could also provide sun protection.

Safety goggles

The importance of using safety goggles can't be stressed enough, especially when connecting your battery and inverter. Always use safety goggles when working on these.

Now that you know the basics of electricity and know which tools you need let's continue by listing the equipment you need.

Equipment

Here is a list of the equipment you need when building a solar system.

Wire Lugs

Wire lugs are tin-plated copper required to make a solid connection in your system. Multiple options are available from different manufacturers and materials. You should only use tin-plated copper. Copper is corrosive, and the tin layer protects it from corroding.

Make sure that the hole fits the battery terminal or inverter connections. You will need a tool to crimp the wire lugs onto the wire, which we discussed in the tools chapter.

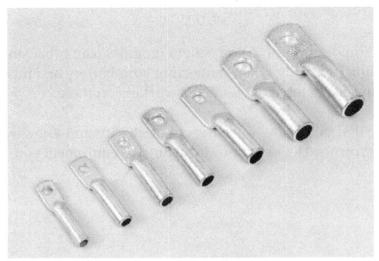

Wire lugs for different AWG wire

You can buy battery and inverter cables with the terminal lugs attached. This way, you don't need special tools to fit these wire lugs. Remember that the thickness of the wire depends on the current that has to flow through it. It is recommended that you buy prefabricated wires if you are using big lugs.

Adding heat shrink over the lugs strengthens the connection, stiffening the wire ends. This reduces the chance of breaking the individual wire strands in sharp bends.

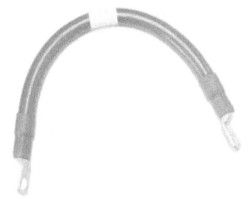

Example of a cable with wire lugs and heat shrink.

Wire lugs are used only for big cables. Crimp connectors are used for smaller wires. Look for UL-listed wire lugs because cheap lugs might not have thick enough material. Your local hardware store will have these.

If you buy lugs, look for lugs that cost $1 a piece minimum. Cheaper ones will have thin walls, making a bad connection between the cable and the lug.

I made a short video about this on my YouTube channel. Check it out here: https://cleversolarpower.com/book/videolugs

Crimp Connectors

As we will discuss in the wiring chapter later, you need to use stranded wire. The only downside to using stranded wire is that you need crimp connectors at both ends to connect your terminals to other devices.

Using crimp connectors gives a better point of contact to the device terminals, reducing heat loss. It will also eliminate corrosion at the exposed sides of the stranded wire.

There are several types of crimp connectors:

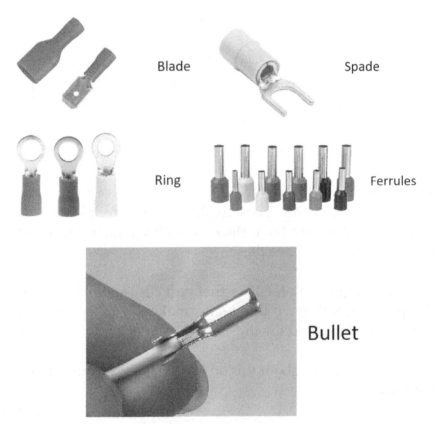

Blade

Spade

Ring

Ferrules

Bullet

Several types of crimp connectors

Ferrules, rings, and spades are the most common connectors in solar applications. Ferrules connect devices' terminals, while rings and spades connect busbars. Bullet connectors are used for MC-4 connectors.

Like most crimp connectors, they come in several colors. Each wire diameter has its own color.

Different-sized ferrule connectors

- AWG 10 (Green)
- AWG 12 (Gray)
- AWG 14 (Blue)
- AWG 16 (Black)
- AWG 18 (Red)
- AWG 20 (White)
- AWG 22 (Orange)
- AWG 22 (Yellow)

Color coding is not always standardized, so they can differ. Once again, it would be best to buy UL-listed wire lugs from your local hardware store.

MC4 connectors

Bullet connectors can be found in MC4 connectors. These are used to bring the electricity from the solar panels to your combiner box or straight to your charge controller. The plastic MC4 connectors protect the cable from moisture, dust, and rain.

They also function as a plug-and-play wiring method for combining solar panels in a string or array (series and parallel). Be careful because these Y-branch connectors can have a maximum of 20-30 amps.

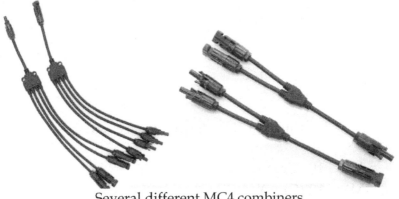

Several different MC4 combiners

Almost all solar panels come with MC4 connectors these days. Here is a guide on crimping them if your panels don't come with them.

After stripping the wire insulation with a wire stripper, you can place the stripped wire in the bullet connector. Use a crimping tool to apply pressure on the crimp connector to secure good conductivity.

Inserting the stripped wire in the bullet connector

Crimping the wire inside the connector

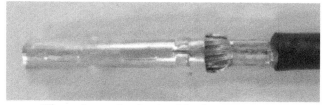

The crimped wire inside the connector

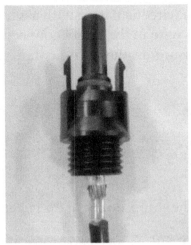

Inserting the crimped connector in the MC4 housing
Source: Marine How-To

Next, you need to tighten the connector to the wire using an assembly tool delivered together with the MC4 connectors.

MC4 Assembly tool

You can buy these cables if you do not want to make them yourself. This is easier and will reduce the possibility of error. Search for 'MC4 connector cables.' Select the correct gauge for the current that flows through it. Parallel connections need bigger wires. More on this later.

Busbars

In electrical power distribution, the busbar is a crucial element to consider in any installation. You need to use a busbar when you use three or more connectors or lugs on any terminal (for example, the battery terminal).

These copper or aluminum strips can be seen inside switchgear or panel boards that carry the currents in the electrical system. They act as the collection or distribution of electrical currents up to the loads from the source. They are also called central wiring terminals. There are several uses for busbars:

- Positive busbar
- Neutral busbar
- Ground busbar

Small busbars are intended for small, off-grid PV applications. They have just a few pins for interconnection between components (inverter, charge controller, and batteries).

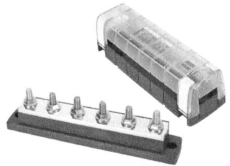

Small 250A busbar

Check the amp rating of a busbar before you buy it. In the 'sizing your solar system' chapter, we will determine the maximum amps in your system.

You can also use fused busbars. These are convenient because you will need to crimp fewer wires.

Victron busbar with integrated fuse holders for MEGA fuses

Displays

To have a visual indication of the charging state of your battery or how much solar power output is generated, you need to have a display instrument. This device will show the values of the variables related to voltage, current, and power that you can put in an easy-to-access location.

One example of such a case is the battery monitor for a battery bank. In the following image, you can see the voltage level of a lead-acid battery. When the battery is under load (when you are using electricity), the battery's voltage will drop. To get an accurate measurement, disconnect or switch off your loads for an accurate voltage reading.

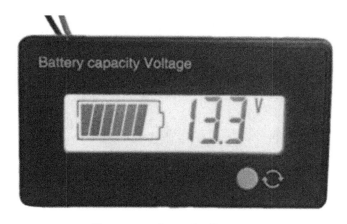

Battery voltage indicator

Many other displays are available, such as an external display from the charge controller or the shunt, which gives you a complete overview of your system. We will discuss these later in the book. Shunts are more accurate and display more information.

Cable Gland

You will most likely need to run your cables through your RV's roof. It's best to use a cable gland for this. To make it watertight, use roof sealant or lap sealant.

Tip: protect your cables from abrasion against the sharp edges of the hole in the roof. Use protective cable sleeves to stop the wires from cutting themselves and making a short circuit.

Double cable entry gland

Most people with off-grid solar in RVs and camper vans will use this cable entry gland.

Depending on the length of the cable, I do not recommend using branch connectors when the combined current is over 20 amps. This is because branch connectors are limited to 10AWG. A 10AWG cable can handle 30 amps, but you also have to take into account the voltage drop to the charge controller. I will discuss sizing wire later in the book.

If the combined current of the parallel setup is more than 20 Amps, you need a combiner box.

Transfer Switch

A transfer switch plays a crucial role in off-grid electrical systems, especially in setups with an alternative power source like a generator or a connection to the utility grid.

A transfer switch is an electrical device that switches a load between two sources. In off-grid systems, it's typically switched between the off-grid power system (like solar panels with batteries) and the primary utility grid or generator.

- **Safety**: The primary function of a transfer switch is to ensure safety. It prevents the backfeeding of electricity into the grid, which can be dangerous to utility workers and can cause damage to your system when grid power is restored.
- **Legal Compliance**: In many regions, a transfer switch is legally required when connecting a generator or an alternative power source to a home's electrical system.
- **Seamless Transition**: Transfer switches allow a seamless transition between power sources. This is particularly important in off-grid systems where reliability and continuity of power are crucial.
- **Load Management**: In some advanced systems, transfer switches can also help manage loads by prioritizing specific circuits or appliances when switching between power sources.

There are two Types of Transfer Switches:

- **Manual Transfer Switches** require manual operation to switch between power sources. They are more straightforward and usually more affordable but require physical intervention.
- **Automatic Transfer Switches (ATS)**: automatically transfer power sources when they detect a loss or restoration of power. They are ideal for ensuring continuous power supply without manual intervention. They can vary in response time. The quick ones will not interrupt the power to your computer or TV, so it keeps working. Others are slower, and your devices will lose power for some time, usually less than a second.

Application in Off-Grid Systems:

- **With Backup Generators**: In off-grid systems with a backup generator, the transfer switch will automatically or manually switch to generator power when the batteries are low or during maintenance.

- **Grid-Interactive Systems**: For off-grid systems with a grid connection for backup or feed-in purposes, like a van on a campground or a boat in a harbor, a transfer switch ensures that the system operates safely from shore power or the batteries.

Under normal conditions, the inverter powers the AC switchboard. It gets its power from the battery. When the battery is depleted, the inverter turns off. The automatic transfer switch senses this and switches the power source from the inverter to the grid or generator.

The speed at which this happens is essential. This should be under 20ms for your computer or TV to keep working without shutting down.

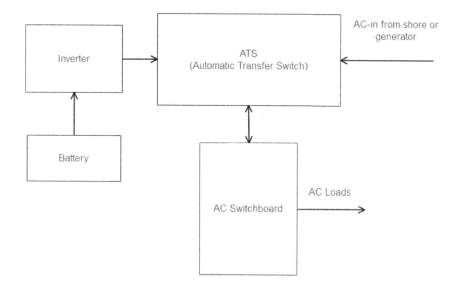

Simplified diagram of a transfer switch

A popular ATS device for vans is one from MOES. It is suitable for 12V,24V, and 48V systems and is limited to 50A or 80A for the upgraded version. The voltage ranges for the switchover is programmable.

MOES transfer switch

Combiner box

This component is a box that contains all the connections coming from every string of solar panels and joins them in a single wire. It is used when connecting more than a few panels in parallel.

From this connection, three larger wire gauge output cables (positive, negative, and ground) contain all the generated DC electricity and transport it to the charge controller. The combiner box consists of a negative bus bar, a ground bus bar, a positive bus bar, circuit breakers or fuses, and an optional surge protection device.

The combiner box is usually set as close as possible to the string of PV modules to reduce voltage drop or DC wiring ohmic losses. Therefore, in residential or commercial applications, they are typically placed outdoors, depending on the type of PV system. Refer to the mounting instructions of the combiner box if you will use one.

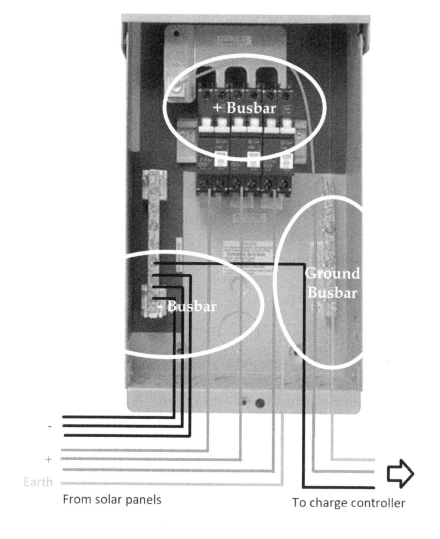

Combiner Box Wiring scheme
Based on the MidNite 20A combiner box

For smaller systems, such as an RV or van, you can use the branch connectors or an outdoor junction box with cable glands.

Small junction box

When selecting a combiner box, you must be aware of several factors:

Encapsulating Rating

Typical encapsulating ratings are classified under the National Electrical Manufacturers Association (NEMA) standards. Type 3R and type 4X are typical encapsulating ratings for combiner boxes.

The type 3R rating enclosure is constructed for either indoor or outdoor use. It protects the equipment inside against incoming solid particles (dirt) and water ingress (rain, sleet, or snow). It also protects the equipment against ice formation on the exterior side.

Meanwhile, the type 4X enclosure-rated combiner box protects all internal equipment from windblown dust and water ingress (rain, sleet, snow, or splashing water). It also protects against corrosion and ice formation on the exterior side. Choose IP65 or higher if you want a waterproof box.

Multiple solar cables should be fed using a PV wire cord grip. This plug takes several PV wires and makes the enclosure watertight. You can also use single cable glands.

PV wire cord grip

The output wire, which goes to your charge controller, needs regular cable glands.

Maximum Voltage Capacity

The combiner box is also designed to withstand a specific voltage rating to provide insulation. Typical low-voltage applications for off-grid purposes will be rated at 600VDC.

Fuse or Breaker Capacity

Also, the combiner box will generally have a specific rating for fuses and breakers in volts. The number of breakers/fuses that can be placed inside is essential, indicating if the combiner box can connect all the PV strings.

Before we end the equipment chapter, let's discuss fuses and circuit breakers. They are vital to every installation.

Fuses and Circuit Breakers

Fuses or circuit breakers in your solar system are not intended to protect the device it is wired to.

We use fuses or circuit breakers to protect the system's wiring from getting hot, melting, or even catching fire.

Therefore, the fuses or circuit breakers are calculated based on the wiring. This protects your system from catching fire if a higher current flows through the wires for which they are rated. This is how you determine fuse sizes:

1. Figure out the load.
2. Decide fuse rating
3. Decide wire thickness based on fuse rating

There is an exception to wires that come directly from solar panels. The wiring coming from PV panels is larger than it needs to minimize the voltage drop. The back of the solar panel will display the maximum allowed fuse size (more on this later).

An example is that you will be running bigger wires rated for 30 Amps to minimize voltage drop, but the maximum fuse for the solar panels is only 10 Amps. We will discuss voltage drop in the wiring chapter.

First Solar, LLC
28101 Cedar Park Blvd.
Perrysburg, OH 43551
First Solar.   www.FirstSolar.com

CE

TUV SKII

Made in USA

Nominal Power (+/-5 %)	60 W
Current at mpp	0.97 A
Voltage at mpp	62 V
Short Circuit Current	1.15 A
Open Circuit Voltage	90 V
Maximum System Voltage	(600V UL) 1000 V
Max Source Circuit Fuse	(2A UL) 10 A
Protection Class	Class II
Cell Type	CdTe

specifications (+/- 10 %) at STC: Irad. 1,000 W/m², AM 1.5, Cell T 25 °C

Technical data of a solar panel

Where to Place Fuses

Fuses should be placed as close as possible to the energy source. Current flows from your battery to your inverter, so put the fuse as close to the battery as possible. Current flows from solar panels to the charge controller, so place them close to the solar panels.

Only place fuses on the positive (red) wire.

Fuses should be put in the following locations:

- You can use inline MC-4 fuses or use fuses in a combiner box when you decide to wire in parallel. (More about series and parallel fusing of solar panels later).

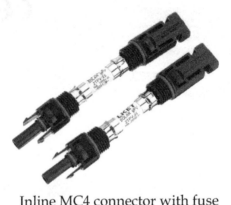

Inline MC4 connector with fuse

- On the positive wire from the charge controller to the battery.
- On the positive wire from the battery to the busbar.
- One the positive wire from the busbar to the inverter.
- On the positive wire from the busbar to the DC loads.

The following drawing will show you where you need to place your fuses.

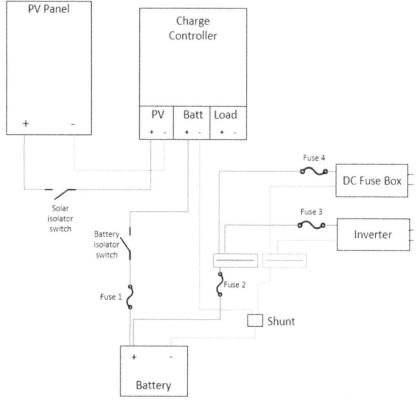

Placement of fuses in your solar system

Fuses vs. Circuit Breakers

DC protection devices are essential to guarantee any PV system's safe and effective functioning and operation. Always make sure you are using DC fuses in your DC system. When you break DC power, there is a larger arc than AC power. It's easier to break AC power because it passes through zero 50 or 60 times per second, while DC doesn't. Therefore, DC protection devices are made differently than AC protection devices.

There are two main types of overcurrent protection devices: Fuses and circuit breakers.

Fuses contain a filament inside that heats up as current flows through it. When a specific current above the permissible limit passes through the filament, it heats up above its thermal capacity and melts. When the wire inside the fuse melts, the circuit gets opened.

An overcurrent can be created by:

- An overload caused by excessive current.
- A short circuit caused by a fault.

Fuse Holder

Fuse

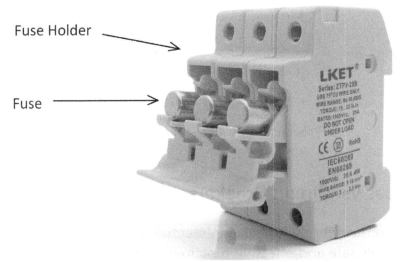

Fuse holder with removable fuses

On the other hand, the circuit breaker is another popular protection device intended for overcurrent protection.

A thermal protection mechanism on a bimetallic contact heats and expands when an electric current above the rated value is present, protecting the circuit against overload. A magnetic protection mechanism instantly responds to high fault currents, protecting the electrical circuit against short-circuits or overcurrents.

Two contacts split inside the DC breaker when an overcurrent passes through the protection device, automatically switching it to the OFF position. The DC breaker needs to be put back in the ON position to allow the electric current to flow again. If a fault occurs with a fuse, you need to replace it. With a breaker, you flip the switch back in the on position. Fuses are cheaper than circuit breakers.

DC circuit breakers are expensive, so many people choose DC fuses. Keep a few spare fuses with you all the time.

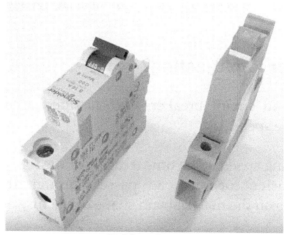
DC breakers

Typically, you use AC breakers after the inverter in your central AC panel. Use fuses for the DC side because they are cheaper.

Do not use a cheap resettable DC breaker like the one in the image. They are known to not break the circuit at the rated current (even at 300% of the current). Always use branded fuses and breakers.

Do not use these resettable DC breakers

Slow or Rapid-acting?

Fuses and circuit breakers are classified according to their response speed.

The acting speed is the time it takes for the fuse to break once a fault current or overload passes through the filament. It depends on the material and shape of the fuse element.

Selecting the correct fuse type involves selecting the proper current rating and response time. A fuse that acts too fast may not allow normal current operations to run, while a fuse that is too slow may not interrupt faulty currents quickly enough. There are mainly three fuse speeds:

- Ultra-rapid
- Fast-acting
- Slow-acting

Ultra-rapid fuses are used for semiconductors (electronics) protection.

Fast-acting fuses protect cabling and less sensitive components such as batteries and PV modules.

Finally, slow-acting fuses feature a built-in delay that temporarily allows the flow of inrush electrical currents for electrical motors.

When checking the datasheet of the fuse, you may find some of the following markings, as described in the following table:

Marking	Description
FF	Very Fast Acting Fuse
F	Fast Acting Fuse
M	Medium Acting Fuse
T	Slow Acting Fuse
TT	Very Slow Acting Fuse

Generally, you will need FF, F, or M-type fuse ratings for battery and solar panel protection. You might need to select a slow-acting fuse to allow starting current to flow if you intend to protect a load like a motor or a pump. More on this in the 'load types chapter'.

Electrical engineers analyze this aspect in detail, considering time vs. current graphs of the fuse, to ensure that the protection device acts when it needs to.

Let's look at different fuses and circuit breakers and where to use them.

Spade Fuses

A type of fuse widely used in solar power applications is the spade fuse, also called blade fuse. These are found in the electrical fuse box of most cars. Their principle is the same as described before. You have to replace them once they trip.

These can be used to act as overcurrent protection for your DC loads.

Six spade fuses

The color of the spade fuses indicates their current rating.

Color	Current
Dark blue	0.5 A
Black	1 A
Gray	2 A
Violet	3 A
Pink	4 A
Tan	5 A
Brown	7.5 A
Red	10 A
Blue	15 A
Yellow	20 A
Clear	25 A
Green	30 A
Blue-green	35 A
Orange	40 A
Red	50 A

Blue	60 A
Amber/tan	70 A
Clear	80 A
Violet	100 A
Purple	120 A

Spade fuses are used in the part of your system with DC loads. Using a fuse box will give you a neatly organized DC load box for LED lights or ceiling fans.

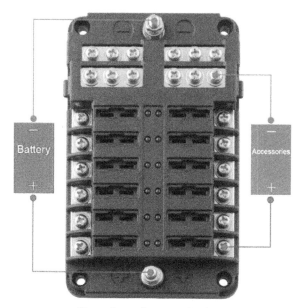

Fuse box for spade fuses

MIDI/AMI

Let's talk about MIDI or AMI fuses. These are specifically designed for systems with a maximum voltage of 32V, which means they're unsuitable for 48V systems. One key specification is their Interrupt Current Capacity (ICC), which is 5000A at 16VDC. This ICC figure is crucial because it indicates the maximum current the fuse can safely interrupt during a short circuit. I will talk more about it at the end of the video. This fuse cost around $15.

These fuses are available from 40 to 200A
For installation, you'll need a specific fuse holder. The fuse holder uses M5 studs, so you'll need an M6 lug for a proper fit. You can also use a fuse block with several fuse holders in one. These fuses are generally a great choice for smaller systems, providing reliable protection in a compact form.

There are also MIDI fuses rated at 70V, but these are not commonly available.

A MIDI fuse

MEGA/AMG

Moving on to a larger option, these fuses share the same Interrupt Current Capacity (ICC) values as MIDI fuses. I don't recommend using them in 48V systems. Their ICC is 2000A at 32V. These fuses are used in the victron lynx distributor or busbar. These fuses cost around $12 a piece.

The fuses are available from 40 to 500A. Typically, these fuses are used as the main battery fuse, often in combination with MIDI fuses. (show fuse block with mega and midi fuses). They are unsuitable for big lithium batteries, which I will discuss later. For installation, you should use a specific fuse holder designed for these fuses, which requires M8 lugs for proper connection.

There are also 70V MEGA fuses, but these are not commonly available.

MEGA fuse

MEGA fuses are often used in fused busbars, which have the advantage of requiring fewer crimps for cabling.

Victron busbar with MEGA fuses

MRBF

These fuses are designed to mount directly onto battery terminals and are suitable for 12 to 24V battery systems. While they are technically rated for up to 58V, I advise against using them with 48V lithium batteries with more than 200Ah capacity. However, they are compatible with 48V lead-acid batteries because the short circuit current in lead-acid batteries is lower.

The Interrupt Current Capacity for these fuses varies depending on the voltage: 10,000A at 14V, 5,000A at 32V, and 2,000A at 58V. These cost around $20 each.

These fuses are available from 50 to 300A. A key consideration for these fuses is their mounting requirement. They need to be mounted on an M10 terminal, which can be a limitation since most batteries come with M8 terminals. However, you can also find them in M8 variants or use a large washer. Their use is generally recommended for lead-acid batteries of up to 48V.

MRBF

ANL

These fuses are well-suited for 12 and 24V batteries, with a maximum voltage rating of 32V. It's important to note that they should not be used for 48V systems. The Interrupt Current Capacity for these fuses is 6000A at 32V. These cost about $20 each.

These fuses are available from 35 to 750A. An interesting feature of the fuse holder is the ability to break the circuit by loosening the bolts and turning the fuse outward. However, this should not be relied upon as a primary safety disconnect, as a small bump could potentially complete the circuit again. Typically, the lugs for the holder are M8, which is a standard size for many fuse holders.

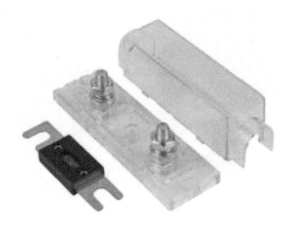

ANL fuse with case

I don't use ANL fuses. They are larger than MIDI and MEGA fuses, and I don't need these high-current fuses.

Class-T

Class-T fuses are the go-to choice for larger 12V, 24V, and 48V lithium battery systems. They stand out with their high Interrupt Current Capacity of 20,000A at 125VDC, making them exceptionally reliable for high-demand applications. These cost about $40 each.

These fuses are available from 100 to 400A. It's important to note that Class-T fuse holders tend to be more expensive compared to other types. You can expect to pay around $50 for a Class-T fuse holder.

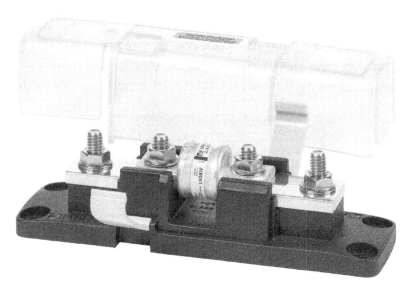

Class-T fuse

NH00

Finally, let's discuss the NH00 fuse, a popular choice in Europe. I frequently used these fuses as an electrician in a chemical factory. It's worth noting that NH00 fuses are not as commonly available in the US.

These fuses are comparable to Class-T fuses but have an even higher Interrupt Current Capacity of 120,000A at 250VDC. When purchasing these fuses, it's crucial to ensure they are DC-rated. They cost around $10 each in Europe. You also need a fuse holder, which will cost around $20.

These fuses are available from 35 to 400A. One significant advantage of NH00 fuses is their cost-effectiveness compared to Class-T fuses.

If they are available in your region, they are a great choice. The fuse holders for NH00 fuses are robust yet not expensive. You can remove the fuse with a special tool or even by hand; I recommend using the tool for safety, especially in 48V systems.

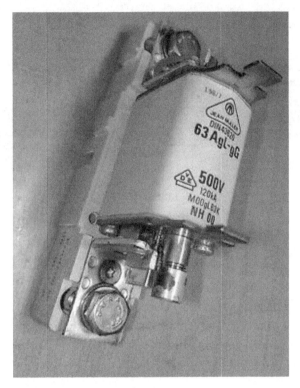

NH00 fuse in its holder

Next, let's talk about why the interrupt current capacity is important:

ICC

A battery fuse's ICC (short circuit current) is a critical safety feature, especially in overcurrent situations. Lithium batteries can deliver a high current very quickly, and in events like short circuits, the current can escalate rapidly to dangerous levels. This is because lithium has a low internal resistance.

The ICC of a fuse marks its maximum safe interrupting current, ensuring it can effectively break the circuit under such extreme conditions without failing. This current is often significantly higher than the battery's nominal capacity – up to 10 times greater.

For instance, in a system with a 100Ah lithium battery, the ICC could be as high as 1000A. This high ICC rating is essential to safely handle potential surge currents and protect the system from fire, battery damage, or more severe hazards.

The higher the battery voltage, the lower the ICC of the fuse. Not only is the ICC important, but so is the battery voltage.

Battery voltage

Selecting a fuse with the appropriate voltage rating for a battery system is crucial, mainly to prevent arcing. Arcing occurs when a fuse blows and an electrical discharge bridges the gap in the circuit. This is particularly concerning in high-voltage systems, such as 48V lithium batteries. A correctly rated fuse ensures that any arc formed is quickly extinguished, preventing potential damage or safety hazards. class-t fuses and NH00 fuses are encapsulated with sand. This sand will extinguish the arc. That's why they have such a high current interrupt rating.

Here is a table which will make your choices for fuses a bit easier:

Fuse Type	Battery Voltage	Current Range	ICC Rating	Fuse Cost	Holder Cost
MIDI	12 and 24V	30-200A	2000A@32V	$15	$10
MEGA	12 and 24V	40-500A	2000A@32V	$12	$15
MRBF	12, 24, 48V (Lead-acid)	50-300A	2000A@58V	$20	N/A
ANL	12 and 24V	35-500A	6000A@32V	$20	$16
Class T	Up to 48V (High capacity lithium)	100-400A	20,000A@125V	$40	$50
NH00	Up to 48V (High capacity lithium)	35-400A	120,000A@250V	$10	$12

Selecting a fuse based on voltage and ICC

If you have a 48V Lithium battery at 300Ah, the short circuit current of this battery is:

$$300Ah \times 10 = 3,000A$$

From the table, you can see that we cannot use an MRBF because it's only rated at 2000A Icc. We will have to use a class-T or NH00 fuse.

Recommended brands

Be cautious about using low-cost fuses and breakers, as they often have reliability issues, which I have shown in one of my YouTube video where I tested MEGA fuses:

https://cleversolarpower.com/book/megatest

Common problems include breaking the circuit too early or failing to break the circuit when needed.

To ensure safety and efficiency in your installation, I recommend choosing breakers from reputable brands known for their quality and testing. Some of the top brands to consider are:

- Blue Sea Systems
- Littelfuse
- Bussman
- Siemens
- Schneider
- Victron

You will pay a few dollars extra, but the safety and peace of mind is invaluable.

Circuit Breakers

As mentioned, fuses are better suited for high-current DC systems, and circuit breakers are used in AC systems. There are mainly three types of circuit breakers:

- Single pole
- Double pole
- Triple pole

Single-pole models are suitable for most circuitry. Simple loads such as fans, TVs, microwaves, coffee makers, home theater equipment, and any other load that works on 120VAC will need a single-pole breaker.

Other loads, such as air conditioners, washing machines, dryers, and some motors, work in split-phase configurations, requiring 240VAC double-pole circuit breakers. Finally, some loads work on three-phase systems at 208VAC, requiring a triple-pole circuit breaker.

Triple pole circuit breakers are not used much in off-grid systems.

A single, double, and triple pole circuit breaker

DC Isolator Switch

DC isolator switches decouple parts of the solar system from each other. You will use them during maintenance tasks.

DC isolator switches are put in these locations:

- Decoupling solar panels from the charge controller.
- Decoupling batteries from the system.

Before buying a DC isolator switch, ensure it complies with the system's current and voltage.

For example, the DC isolator switch (solar disconnect switch) from your solar panels to the charge controller has a lower current but higher voltage.

800 Volts DC, 25 Amp solar disconnect switch

You can also use a double-pole breaker instead of an isolator switch. This is cheaper, but it's not waterproof unless installed in a waterproof box. If you use the breaker instead of the solar disconnect, you can install it right next to your charge controller on a DIN-rail mount.

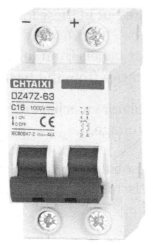

16A DC breaker

The battery disconnect shown is from Blue Sea Systems. Other battery switches have a higher internal resistance and produce much more heat, resulting in undesirable system losses.

In comparison, the isolator switch from the battery requires a higher current but lower voltage (depending on the voltage of your battery bank).

48 Volts DC, 300 Amp battery isolator switch
(blue sea systems)

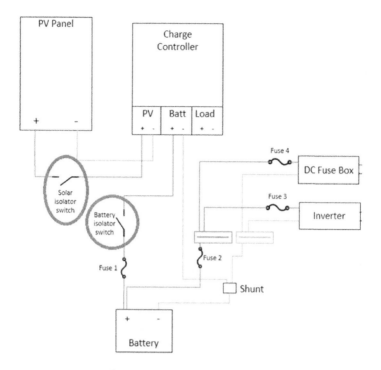

Where to place isolator switches

Basic Circuitry

Basic Light Circuit

A light circuit is one of the most basic electrical circuits. It requires a battery (power source), a switch (control device), and a lightbulb (load).

Placing the battery with the positive terminal toward the switch and then connecting the bulb between the switch and the battery will allow you to manually activate or deactivate the light.

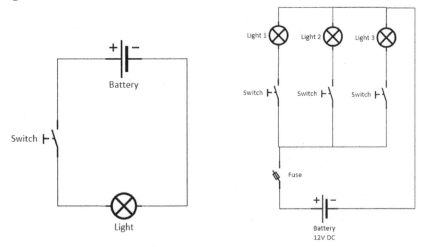

Basic light circuits

It is a good practice to draw out your system before purchasing the components. You don't need a professional computer drawing; you can do it with pen and paper.

Short Circuit

A short circuit is a non-normal operational behavior of an electrical circuit in which a large amount of current flows through an unintended path with low resistance.

Short-circuits are generally associated with two scenarios:

- Short-circuits between phases.
- Short-circuits to ground.

A short circuit occurs when direct contact (metal to metal) occurs between battery terminals or cables with different polarities.

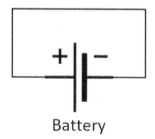

Battery

Short circuit between terminals on a battery

This is unlikely to happen unless you connect the positive and negative from the battery with a single wire (do not try!). Another reason this can occur is if you drop a tool on the battery terminals. Therefore, you should buy the insulated tools described in the 'tools' chapter.

The next chapter is about the different load types in an off-grid system. Some of these tips will help you design your system.

Load Types

An important step when sizing your PV system is to estimate the load that you will have. This topic covers mainly two aspects: power and energy. An off-grid PV system works as a variable and limited energy source.

Therefore, you must determine the electrical loads that will consume power in the PV system and specify how long they will be used.

In this matter, it is essential to know the type of loads you will be connecting to the PV system and understand how they work. Let's see some typical loads you will use for off-grid purposes.

Resistive Loads

There are mainly two types of resistive loads:

- Linear loads
- Non-linear loads

Linear-type loads consume an average amount of power that is constant over time. There are no significant fluctuations while running or starting them. They are generally associated with the behavior of electrical resistance; therefore, they are called resistive loads. Examples include a lightbulb and a water heater.

Non-linear loads behave like inductors or capacitors, which have a variable consumption over time. An example of a non-linear resistive load is a computer with a switching power supply. The power consumption of resistive loads is listed in their datasheets, and it can be expressed in watts or amps.

You will typically find the nominal voltage of the load in the datasheet as well. Using the power formula:

$Watts = Voltage \times Amps$, You will be able to find the equivalent power that the load consumes. The following table shows some resistive loads used for residential purposes.

Appliance	Running Watts
CD/DVD Player	35-100
Clock Radio	10-50
Desktop Computer	60-200
Laptop	20-150
Printer	30-50
Coffee Maker	650-1,200
Hair Dryer	1,000
Blender	1,200
Electric Water Heater	1,500
Fan	30-100
Iron	1,000-1,500
Microwave	1200
24" LED TV	40-50
Air conditioner 5000 BTU	500
Electric Stove	2,000
Electric Blanket	200
LED lights	6-20

Typical resistive load consumptions for an RV

As you can see, the loads that consume more power are the ones that:

- Heat the space
- Cool the space
- Generate heat for cooking

These loads must be selected carefully as they will draw a lot of power and energy from the PV system. Elements such as an electric stove, an electric frypan, or a waffle iron should be avoided and replaced by other energy sources like natural gas. Replacing these energy-demanding appliances will significantly reduce the cost of your system.

Inductive Type Loads

Inductive loads draw more current during their start cycle. Therefore, when starting inductive loads, you need to consider the surge current.

Refrigerator

The most important of the inductive loads is the refrigerator. The compressor takes a cool refrigerant liquid and transforms it into a hot refrigerant liquid with a higher pressure to complete the refrigeration cycle. The compressor needs an electric motor that generates movement inside the compressor. This means we will have a power surge while starting.

In the following image, you can see the power consumption of two 20 cubic feet refrigerators.

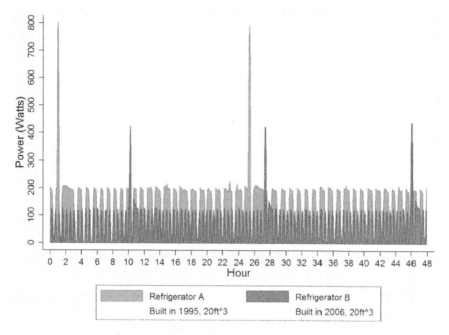

Refrigerator starting pattern old vs. new
Source: Is your refrigerator running?

- Refrigerator A has a nominal power consumption of 200W and a starting power of 800W.
- Refrigerator B has a nominal power consumption of 100W and a starting power of 400W.

In both cases, the starting power is four times the nominal power required for operation. Therefore, the off-grid PV system must always provide this surge power. My 100W fridge cannot run on a 1,000W inverter; it can on a 1,500W inverter. So it doesn't hurt to size the inverter larger than four times the nominal power rating.

Another factor you must consider with refrigerators when sizing the PV system is that you cannot take the nominal power consumption and multiply it by 24 hours. This will lead to oversizing, which is a common mistake.

83

Refrigerator datasheets often include a yellow label indicating the product's energy consumption per year or day. You must use this reference in your calculations for energy yields. Power consumption must also be considered for the inverter's power rating.

I have made a video where I show the daily consumption of my 100W fridge. You can view it using this link: https://cleversolarpower.com/book/fridge

It consumed 0.6kWh per day.

The energy consumption of a fridge depends on many factors:

- **Type of fridge**: A top loader will consume less power than a display fridge.
- **Size**: The volume of the fridge will play a role in energy consumption.
- **Location**: If the fridge is well ventilated at the condenser, it will require less energy.
- **Season**: The fridge needs to work harder during the summer because the temperature difference is higher.
- **Usage frequency**: Opening the door frequently will lead to more energy usage.
- **Temperature set point**: Check to ensure the temperature setpoint is not too cold.
- **Age**: The age of the refrigerator also affects energy usage. The newer, the less energy it will use (if it is the same type).
- **Quality of the seals**: Cold air will leak if the seals are not sealing well.

Depending on all these factors, refrigerators will generally consume 30% of their rated power in one day. 30% of 24hours is 8 hours. For example, a fridge that is rated at 100 Watts and runs for 24 hours a day will consume:

$$100\ Watts\ \times 8\ hours = 800\ Watt\ hours\ per\ day$$

Many people consider running a DC fridge. These are a great option if you have an RV and don't have space for a large fridge. Their price has come down in 2024, so they have become a good option.

My Vevor car fridge consumes 180Wh per day.
Watch my tests: https://cleversolarpower.com/book/carfridge

Washer/Dryer

There are two types of dryers: gas- and electric-based dryers. Electric dryers circulate an electric current through a resistor to generate heat. They consume a considerable amount of electricity, up to 725kWh per year, and a lot of power, reaching over 3kW.

The energy and power required to supply a dryer is too much for an RV or simple off-grid system with a small space available for solar panels.

The best choice might be a gas-based dryer. Using a dryer that works on natural gas comes with other essential safety regulations, like placing it in a well-ventilated place and allowing fresh air to enter the dryer's intake. This can be accomplished by installing an external intake and exhaust pipe. Installing a propane detector is a good safety precaution.

RV Water Pump

An RV water pump is another type of load that you can add to your list. RV water pumps generally work at 12 VDC. They can draw between 2.5 Amps and 10 Amps under regular operation.

3 gallons per minute RV pump from SeaFlo

However, as they also include a DC motor, they could draw between 10 and 40 Amps during the starting process. Since your fuses will be slow-acting, these will not break the circuit when the motor starts (only a few milliseconds).

Keep in mind that these 12V water pumps are designed for intermittent use. In other words, they are intended to be used during the time that you take a shower, wash your hands, or the time it takes you to flush a toilet. These are DC appliances that do not count toward your inverter power.

Air Conditioner

The air conditioner is a convenient but very demanding load. If you are thinking of powering an AC unit with solar panels, you must accurately estimate the energy consumption of this load.

An AC unit's power consumption cannot simply be calculated based on the nominal power. Doing this will represent a tremendous increase in energy demand, and your solar panel system will be oversized.

This device also has a motor that runs a compressor, so it requires a surge current. For AC units, a reasonable assumption is that the surge power will equal three times the electrical power on the technical datasheet.

A common mistake is to assume the air conditioner's energy consumption is related to the number of hours it is used.

You will notice that energy consumption will be much lower. The following image shows the consumption pattern of an AC wall unit to give you a reference to performance behavior.

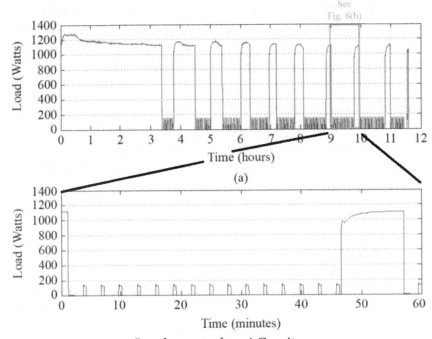

Load curve of an AC unit
Source: Load profiles of selected major household appliances and
their demand response opportunities

As you can see, the AC unit will consume its rated power of
1,200 Watts to cool down the room. After that, the compressor
(outside unit) will stop while only the fan inside will work. The
compressor will have the highest energy consumption.

Energy consumption will greatly depend on the difference in
temperature between the inside and outside, the time of day,
how many times you open the doors, and the insulation, just to
name a few.

In the next chapter, we will apply our knowledge and show you
how to size your system correctly.

Microwave

Microwaves often have two power ratings. The first is the advertised power—how much energy goes directly into heating your food. My microwave is rated at 800 watts in this regard.

On the back, however, is another rating showing the actual power the microwave draws from the outlet, rated at 1,150 watts, though this number can vary.

My 800W microwave draws 1,420 watts—about 1.8 times more than advertised, meaning it's operating at 58% efficiency. This is typical for microwaves. If your microwave is rated at 1,000 watts, it may draw closer to 1,800 watts.

For this reason, I recommend doubling the advertised power rating when calculating your inverter needs. For example, my 800-watt microwave works perfectly with a 2,000-watt inverter.

You might assume that using a lower setting, like 100 watts, means you could get away with a smaller inverter. I tested this to see if it's true. Even on a lower setting, the microwave pulls the same power but cycles on and off.

You still need an inverter that can handle the full load. The exception is inverter microwaves, which can truly adjust their power output to match a lower setting. For instance, setting it to 400 watts means it consistently uses 400 watts, allowing for a smaller inverter, though it takes longer to cook.

Sizing your Solar System

Now that you know the different types of loads and what it takes to run them, we estimate the size of your system to support all your devices.

We will start by briefly describing system voltages. Then, we will work on a practical example.

12, 24, and 48V systems

Knowing the consequences of choosing a specific voltage for your system is essential. We will discuss the advantages and disadvantages of these systems for different components.

Charge controller
Let's say you have a charge controller that can handle a maximum of 80 Amps. This equals a maximum of 960 watts of solar at 12 volts.

$$12V \times 80A = 960\ Watts$$

Let's see what happens when we increase the battery voltage to 24 volts. We now have 1920 watts of solar with the same 80-amp charge controller.

$$24V \times 80A = 1,920\ Watts$$

Or even more when we have a 48-volt battery system.

$$48V \times 80A = 3,840\ Watts$$

The conclusion here is that your charge controller will be cheaper when you use a higher-voltage system. If you increase your battery voltage, you can have a lower-amp charge controller for the same PV input power. The following image illustrates this.

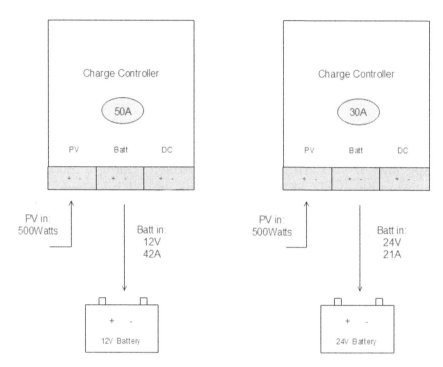

Influence of battery voltage on the charge controller

Wiring
Remember the power formula? We can only increase the current if the voltage is fixed to achieve our desired power output. That means that if the voltage is 12 volts, we must increase the current to a certain level to meet our power requirements.

$$Power = Current \times Voltage$$

If we have a 2,000-watt inverter, we can calculate the current going through the wire at 12 Volts, 24 volts, and 48 volts.

$$Current = \frac{2,000\ watts}{12\ volts} = 167 Amps$$

$$Current = \frac{2,000\ watts}{24\ volts} = 83\ Amps$$

$$Current = \frac{2,000\ watts}{48\ volts} = 42\ Amps$$

Using a 2,000-watt inverter on a 12-volt system will draw 167 amps. I recommend staying under 100 Amps in DIY systems, especially if you crimp the wire yourself. Using a 2,000-watt inverter on a 24- or 48-volt system would be much better. Besides being cheaper, it will be safer and easier to work with because you don't need big wires.

I have listed my recommendations for the inverters below.

- 12V battery -> inverter upto 1000W
- 24V battery -> inverter from 1000-2000W
- 48V battery -> inverter from 2000W to 4000W
- More power -> have multiple inverters in parallel

A fully charged 48-volt battery will be higher than 50 volts at the terminals (58.4V). Be aware that a 48-volt system is more dangerous than a 12- or 24-volt system.

I recommend using a 24 or 48-volt battery for most systems. The wires will be cheaper, and the cost of the charge controller will be more affordable as well. Be careful with lithium batteries. Some cannot connect more than two in series to make a 48 volt system. This is because the battery management system (MOSFETs) is incompatible with the higher voltage. More on this later in the battery section.

Now that you know the benefits of a higher-voltage battery bank, let's calculate the rest of the system.

Calculations

If you would like this in video format, I recommend watching my video here: https://cleversolarpower.com/book/sizing

Let's say you are converting a van to an off-grid mobile home. You are planning to use the following devices:

- Phone charger
- Laptop charger
- Water pump
- 5 DC led lights
- Speaker system
- 12V top-loading fridge/freezer combo
- 12V ceiling fan
- Blender
- Egg cooker

Now, you need to separate the AC devices from the DC devices. Try to do this in a spreadsheet. You can also use my load analysis tool on my website: https://cleversolarpower.com/off-grid-solar-calculator/

Everything with a standard household plug will be an AC device. Anything that works on DC goes into the list of DC devices.

Try to use DC devices instead of AC because it will limit the load on the inverter. That way, you can choose an inverter with a lower power rating.

I will enter my devices in the list and separate them by DC and AC.

DC appliances			
Item			
5 Led lights			
Phone charger			
Water pump			
Speaker system			
Fridge			
Ceiling fan			
AC appliances			
Item			
Laptop charger			
Blender			
Egg cooker			

Categorizing devices into DC and AC

Next, you are going to determine the power rating for each device. There are four options to find the power of a device:

1. By looking at the sticker.

Start by searching for the sticker on the appliance you want to use. If you are lucky, you might find the power (watts) the device consumes. If there is no visible sticker on the device, move on to option two.

Sticker on the device

2. By searching online.

Search online if you have not bought the device yet or can't find the sticker. Search for the item name in Google. You will find some websites that list the power rating on their product page.

3. By using the 'kill a watt meter.'

If you cannot find the device's power rating, you can use a 'kill a watt' meter to read its power consumption. This is especially useful for devices not constantly in use, like a fridge or an air conditioning unit.

You can also use a smart plug. These usually have an app that allows you to monitor power consumption. I use these to determine the power consumption of my fridge.

The 'kill a watt' meter

4. By applying the power formula.

The last method is to search for the device's current. Try to locate the device's current using methods one or two.

$$Power = Voltage \times Current$$

Now, we know the device's voltage, either 12 Volts DC or 120 Volts AC, and the current through it. We can determine the device's power rating.

Let's take this example and calculate the power it consumes.

$$Power = 120\,Volts \times 11.5\,Amps = 1{,}380\,Watts$$

This would be different if we were using a DC device. Let's take a 12VDC water pump which uses 3,3amps.

$$Power = 12\,Volts \times 3.3\,Amps = 40\,Watts$$

Now we fill in the power of each device in the spreadsheet.

DC appliances	
Item	**Watts**
5 Led lights	25
Phone charger	24
Water pump	40
Speaker system	25
Fridge	45
Ceiling fan	25
AC appliances	
Item	**Watts**
Laptop charger	50
Blender	1000
Egg cooker	500

Enter the power rating on the spreadsheet

Now, you can decide the power rating of your inverter. If you do not use your blender and egg cooker together, you can use a 1,500-watt inverter. You should not use a 1,000-watt inverter because it would heat up, reducing its efficiency. It will also make more noise because the cooling fan will be running more often.

The next step is to determine how long you will use these devices each day in hours. Use the following formula to convert minutes to hours:

$$number\ of\ minutes\ \times \left(\frac{1}{60}\right)$$

I'm using a blender for two minutes each day.

$$2\ minutes\ \times \left(\frac{1}{60}\right) = 0.033\ hours$$

Put the time values in the spreadsheet.

DC appliances			
Item	Watts	Hours per day	
5 Led lights	25	4	
Phone charger	24	1.5	
Water pump	40	0.5	
Speaker system	25	4	
Fridge	45	4	
Ceiling fan	25	1	
AC appliances			
Item	Watts	Hours per day	
Laptop charger	50	4	
Blender	1000	0.03	
Egg cooker	500	0.16	

Calculating hours per day

In this step, we will calculate the number of watt-hours the devices consume in one day. We do this by using the following formula, which we have already learned.

$$Watt\ hours = Watts \times hours\ per\ day$$

For the 5 led lights, this is:

$$25\ Watts \times 4\ hours = 100\ Watt\ hours$$

DC appliances			
Item	Watts	Hours per day	Watt hours
5 Led lights	25	4	100
Phone charger	24	1.5	36
Water pump	40	0.5	20
Speaker system	25	4	100
Fridge	45	4	180
Ceiling fan	25	1	25
		Total Watt hours DC	461
AC appliances			
Item	Watts	Hours per day	Watt hours
Laptop charger	50	4	200
Blender	1000	0.03	30
Egg cooker	500	0.16	80
		Total Watt hours AC	310
	Total Watt hours per day:		771

Calculating the total watt-hours for our system

Now we can add up the total DC and AC Watt-hours.

The next step is to calculate the battery we will need. Battery capacity is expressed in amp-hours or Ah.

Now, you need to decide the battery bank voltage. Remember the previous chapter about 12, 24, and 48-volt systems? We are going to limit the current in the wires to a maximum of 100 Amps. Now, you can select a voltage level based on your inverter size.

For this example, we will continue using 12 volts for the battery bank. Next, we calculate the battery's Amp-hours need. The value we get from our load-analysis table is 771 watt-hours.

$$Amp\ hours = \frac{Watt\ hours}{Voltage}$$

$$\frac{771\ Watt\ hours}{12\ Volts} = 65Ah$$

If you use a lead-acid battery, you need to double the capacity because it can only be discharged to 50%. A 100Ah lead-acid battery only has 50Ah of usable energy. You can get 50% of usable energy out of the lead-acid battery.

Lithium batteries can be cycled from 10 to 90%. A 100Ah lithium battery has 80Ah usable energy, so you can get 80% of usable energy out of it. While you technically can get 100% usable capacity, using only 80% will be conservative (more on this later).

A requirement of 65Ah per day will need a battery with the capacity of:

For lead-acid:

$$65Ah \times \left(\frac{100\%}{50\%\ usable\ capacity}\right) = 130Ah$$

For lithium:

$$65Ah \times \left(\frac{100\%}{80\% \ usable \ capacity}\right) = 81.25Ah$$

Unfortunately, there will be some days of shade where your batteries will not be fully charged. This will depend on:

- The place you have your setup (latitude).
- Time of year (summer or winter).
- Weather (cloudy or sunny).

Having at least two days of autonomy is recommended, but three days is better. Let's use three days. This means that you need the following battery size:

For lead-acid:

$$130Ah \times 3 \ days = 390Ah$$

For lithium:

$$81.25Ah \times 3 \ days = 243.75Ah$$

Next, you need to consider the efficiency of the type of battery you will use. Here are the efficiencies of various batteries:

- Lead-acid: 80%
- AGM: 90%
- Lithium: 99%

For lead-acid:

$$390Ah \times \left(\frac{100\%}{80\%}\right) = 487.5Ah$$

For Lithium:

$$243.75Ah \times \left(\frac{100\%}{99\%}\right) = 246.2Ah$$

If you use a lead-acid battery, you need a battery of 487.5Ah at 12 Volts.

If you are using a lithium battery, you need a battery of 246.2Ah at 12 Volts.

The next step is to determine the recommended power of the solar panels to charge the battery in one day. To do this, we need to convert our battery bank size from amp-hours to watt-hours. We have chosen a 12 Volt battery system, so we will multiply the amp-hours by 12 Volts.

$$Watt\ hours = Voltage \times Amp\ hours$$

For lead-acid:

$$12\ Volts \times 487.5Ah = 5,850\ Watt\ hours$$

For lithium:

$$12\ Volts \times 246.2Ah = 2,954\ Watt\ hours$$

Remember that you can use 50% of the energy for a lead-acid battery and 80% for a lithium battery? That means there will always be 50% for lead-acid and 20% for lithium left in the battery. That's why these values are different. Let's calculate the usable watt-hours that can be stored in these batteries.

For lead-acid:

$$5{,}850 \; Wh \times 50\% = 2{,}925 \; Wh$$

For lithium:

$$2{,}954 \; Wh \times 80\% = 2{,}363 \; Wh$$

The amount of watt-hours we have just calculated is what we have to recharge in one day with solar power. You can also recharge the battery with a combination of shore, generator, or alternator charging. But for the sake of this exercise, let's stick with solar power alone.

Now we have to figure out the amount of sunhours our location gets during the year.

We will use a website called PVWatts. The tool is free to use here: https://pvwatts.nrel.gov. It can be used anywhere on earth.

Enter a location in the search bar and press go. In this case, I chose Houston, Texas.

You will then see this page:

RESOURCE DATA   SYSTEM INFO   RESULTS

SOLAR RESOURCE DATA

The latitude and longitude of the solar resource data site is shown below, along with the distance between your location and the center of the site grid cell. Use this data unless you have a reason to change it.

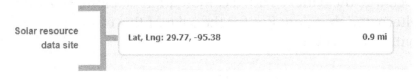

Check if the location is correct on the map underneath (it is not shown in this image) and note down the latitude. In this case, it's 29.77 degrees. We need this to calculate the ideal solar panel tilt later.

As you will soon learn in the solar panel chapter, to get the most solar power throughout the year, you must use the year-round tilt.

You must take your latitude and subtract it by 2.5 degrees. So, for Houston this becomes:

$$29.77° - 2.5° = 27.27°$$

If you don't like to calculate this yourself, you can use my calculator on my website:

https://cleversolarpower.com/book/tilt

Then, click on 'go to systems info'.

You will now see the following page:

SYSTEM INFO

Modify the inputs below to run the simulation.

DC System Size (kW):	1	ⓘ
Module Type:	Standard	ⓘ
Array Type:	Fixed (open rack)	ⓘ
System Losses (%):	14.08	ⓘ
Tilt (deg):	27.7	ⓘ
Azimuth (deg):	180	ⓘ

- Enter the size of your system, let's use 1kw for now. We will come back and adjust it later.
- Select module type standard.
- Select fixed, ground mount, or a tracker.
- The system losses are standard, so there is no need to adjust that.
- Then we get to the tilt degrees. Enter the values we just calculated which is 27.7°.
- The azimuth is 180 for the northern hemisphere and 0 degrees for the southern hemisphere.

Then click on 'go to pv watts results'

Then we get to see these results:

Month	Solar Radiation (kWh / m^2 / day)
January	3.99
February	4.13
March	5.44
April	5.45
May	5.74
June	5.83
July	5.94
August	5.82
September	5.60
October	5.43
November	4.95
December	3.70

You will see the month on the left and the solar radiation or sunhours per day on the right.

"Sunhours," or "peak sun hours," is a term used in solar power to measure the total amount of solar energy available to generate electricity in a given location, averaged daily. Rather than referring to the actual daylight hours, sunhours quantify the intensity and duration of sunlight in terms of how many hours of full, direct sunlight (or 1,000 watts per square meter) a solar panel would receive each day.

For example, a region with 5 peak sun hours means that, on average, the area receives enough solar irradiance equivalent to 5 hours of direct sunlight daily. This metric is essential for sizing solar systems accurately since it reflects the power production potential based on local solar conditions.

You might ask, 'I have 10 hours of sun daily. Why do you say I have 5 sun hours? Here's why:

One sun hour equals $1,000W/m^2$ of solar irradiance. So, 10 hours at $500W/m^2$ equals 5 sun hours. Daylight hours aren't the same as sun hours."

You will see that it's lower in winter than in summer. The lowest is 3.7 in December, and the highest is 5.94 in July. We can use this number to determine the number of solar panels we need.

Remember we must recharge our lead acid battery with 2,925Wh and lithium with 2,363Wh?

For lead acid:
$$\frac{2,925Wh}{3.7\ sunhours} = 790W$$

And for lithium:

$$\frac{2,363Wh}{3.7\ sunhours} = 638W$$

Now, we don't need to use a minimum of 3.7 sunhours. We can run the generator for three months to charge the battery during the winter. So we have to calculate the power of solar panels based on 5 sunhours.

For lead acid:

$$\frac{2,925Wh}{5\ sunhours} = 585W$$

And for lithium:

$$\frac{2,363Wh}{5\ sunhours} = 472W$$

We can see that we need fewer solar panels. This might seem insignificant, but the difference becomes greater during summer and winter when you go further north. And it might be more cost-effective to run the generator in the winter months.

Choose which solar panel you will use. In this example, we will use 100-watt panels.

For lead-acid:

$$\frac{790\ Watts}{100\ Watt\ panel} = 8\ Panels$$

You need eight panels of 100 Watts each to charge your 12V lead-acid battery bank in one day.

For lithium:

$$\frac{638\ Watts}{100\ Watt\ panel} = 7\ Panels$$

You need seven panels of 100 Watts each to charge your 12V lithium battery bank in one day.

Now that you know how many solar panels you will get, we can run a full simulation on the PVwatts website.

Let's say you are going to use lithium batteries. We have to adjust the panels' power to 700 Watts or 0.7.

SYSTEM INFO

Modify the inputs below to run the simulation.

DC System Size (kW):	0.7
Module Type:	Standard
Array Type:	Fixed (open rack)
System Losses (%):	14.08
Tilt (deg):	27.7
Azimuth (deg):	180

- 0.7 or 700W of solar power.
- We have a standard solar panel module.
- We have a fixed installation for our solar panels, which means they are on a rack on the ground.
- We have efficiency losses of 14.08% (standard).
- We have adjusted the tilt angle to 27.7 degrees for maximum solar energy harvest.
- And we will point the panels south (180 degrees). If we were in the southern hemisphere (below the equator), we would have to point the panels north, which means an azimuth of 0 degrees.

These are the expected results for your system throughout the year:

Month	Solar Radiation (kWh / m^2 / day)	AC Energy (kWh)
January	3.99	69
February	4.13	64
March	5.44	91
April	5.45	87
May	5.74	93
June	5.83	90
July	5.94	94
August	5.82	92
September	5.60	87
October	5.43	89
November	4.95	81
December	3.70	65
Annual	5.17	1,002

Remember, our daily consumption is 700Wh. Over 30 days, that adds up to 21,000Wh, or 21kWh. To account for three days of autonomy, we multiply by three, giving us 63kWh of required energy storage for the month. This ensures that even in winter, we can harvest enough energy to meet our needs. And as you can see, we reach this amount every month.

Wiring

Wiring is an essential part of any electrical installation. Solar installations also need special consideration in this matter.

In this chapter, we will look at the wire core material, different types of wires, factors that contribute to wire sizing, and how to calculate the diameter of the wire.

Wire core material

There are many types of wiring you can use. Stranded wire, which consists of multiple wires in one, is recommended. This has the advantage of being flexible, while solid cables are tough to work with. Furthermore, stranded wires are better for DC applications than solid-core wires.

Stranded flexible wires

If you buy wires in the store, you will have three options:

- Copper wire
- Copper-clad aluminum
- Aluminum wire

111

Copper wire is a better conductor than aluminum but is also more expensive. Since copper is a better conductor of electricity than aluminum, you need to increase the diameter of your aluminum wires to account for this — more information about this later.

Most of the wires sold online are copper-clad aluminum (CCA). It means that while you think you are buying copper cables, you get aluminum cables that look like copper cables. People unaware of this purchase the copper-clad aluminum and size their wires based on copper cable. This can be very dangerous! Always use copper wire, and don't save money on your wires.

Sizing Factors

Wires are rated according to their diameter. We will use the American Wire Gauge (AWG) throughout this book. Depending on the selected size, each wire type offers a different current capacity (ampacity) rating.

These are classified in pair numbers that go from #18AWG up to MCM scale cable sizes (which will not be used here). For our solar PV off-grid applications, we will only use cables between #12AWG and 2/0AWG wire sizes.

To size the wire that you need for your PV system, you must consider these factors:

- Current Capacity or Ampacity
- Ambient temperature
- DC voltage drop

Let's take a closer look at what these terms mean.

Current Capacity

The wire manufacturer provides the current or amperage capacity in their datasheet.

The following table offers a reference from the National Electric Code. In the left column, you see the wire sizes, ranging from 14 to 2000. 14 is the smallest diameter while 2000 is the biggest.

The wire core material is visible at the top. It can be copper or aluminum, but copper is recommended.

Below that, it shows the insulation's temperature rating. This can be 60°C (140°F), 75°C (167°F), or 90°C (194°F). Choosing a wire with a higher temperature rating is always better. Under the temperature ratings, you can see the different types of wire. We will talk about these in the next chapter.

The last and most prominent data representation on the table is the conductor's ampacity rating, or the amount of current that can pass through the wire.

A selection of the American AWG wire sizes (not on a real scale)

National Electrical Code
Allowable Ampacities of Insulated Conductors Rated 0-2000 Volts

As Excerpted from the 2002 National Electrical Code

Ampacities of Not More Than Three Current-Carrying Conductors in Raceway, Cable or Earth. Based on Ambient Temperature of 30°C (86°F)

	Copper Conductors			Aluminum Conductors			
	Temperature Rating of Conductor			Temperature Rating of Conductor			
SIZE	60°C	75°C	90°C	60°C	75°C	90°C	SIZE
AWG OR kcmil	TYPES TW UF	TYPES RHW THW THWN THHW XHHW	TYPES RHH RHW-2 THHW THWN-2 XHHW-2 XHH THW-2 THHN USE-2	TYPES TW UF	TYPES RHW THW THWN THHW XHHW	TYPES RHH RHW-2 THHW THWN-2 XHHW-2 XHH THW-2 THHN USE-2	AWG OR kcmil
14**	20	20	25	-	-	-	-
12**	25	25	30	20	20	25	12**
10**	30	35	40	25	30	35	10**
8	40	50	55	30	40	45	8
6	55	65	75	40	50	60	6
4	70	85*	95*	55	65	75	4
3	85	100*	110*	65	75	85	3
2	95	115*	130*	75	90*	100*	2
1	110	130*	150*	85	100*	115*	1
1/0	125	150*	170*	100	120*	135*	1/0
2/0	145	175*	195*	115	135*	150*	2/0
3/0	165	200*	225*	130	155*	175*	3/0
4/0	195	230*	260*	150	180*	205*	4/0
250	215	255*	290*	170	205*	230*	250
300	240	285	320	190	230*	255*	300
350	260	310*	350*	210	250*	280*	350
400	280	335*	380*	225	270	305	400
500	320	380	430	260	310*	350*	500
600	355	420	475	285	340*	385*	600
700	385	460	520	310	375	420	700
750	400	475	535	320	385	435	750
800	410	490	555	330	395	450	800
900	435	520	585	355	425	480	900
1000	455	545	615	375	445	500	1000
1250	495	590	665	405	485	545	1250
1500	520	625	705	435	520	585	1500
1750	545	650	735	455	545	615	1750
2000	560	665	750	470	560	630	2000

Ampacity ratings for multiple insulated conductors.
Source: National Electric Code

To view this chart, go to:
https://cleversolarpower.com/resources

To select the correct wire size, you must estimate the maximum electrical current that will flow through that wire section and then select the wire gauge that withstands that current.

For example, A 1,000 Watt inverter that feeds itself from a 12 volt battery has a current of:

$$Current = \frac{1000\ Watts}{12\ Volts} = 83.3\ Amps$$

If we use THWN-2 copper wire with insulation rated at 194°F (90°C), at an ambient temperature of 83°F (30°C) and use the previous table, we need a 4AWG wire.

SIZE AWG OR kcmil	Copper Conductors			Aluminum Conductors			SIZE AWG OR kcmil
	Temperature Rating of Conductor			Temperature Rating of Conductor			
	60°C	75°C	90°C	60°C	75°C	90°C	
	TYPES TW UF	TYPES RHW THW THWN	TYPES THHW XHHW USE	TYPES RHH RHW-2 XHHW XHHW-2 XHH	TYPES THHW THWN-2 THW-2 THHN USE-2	TYPES TW UF	TYPES RHW THW THWN	TYPES THHW XHHW USE	TYPES RHH R-HW-2 XHHW XHHW-2 XHH	TYPES THHW THWN-2 THW-2 THHN USE-2	
14**	20	20	25	-	-	-	-
12**	25	25	30	20	20	25	12**
10**	30	35	40	25	30	35	10**
8	40	50	55	30	40	45	8
6	55	65	75	40	50	60	6
4	70	85*	95*	55	65	75	4
3	85	100*	110*	65	75	85	3
2	95	115*	130*	75	90*	100*	2

Selecting wire diameter

The tables provided apply to cables rated for 60, 75, and 90°C. Welding cables, however, are rated higher at 105°C (221°F) and aren't included in these tables. Instead, the cable manufacturer specifies their current capacity. For instance, WindyNation welding cables can handle more current than standard cables due to their higher-temperature insulation.

If possible, I recommend using welding cables for your installation, but always check the manufacturer's datasheet to confirm their temperature rating. Be cautious with suppliers who label a cable as 'welding cable,' as many are PVC cables rated only to 90°C (194°F).

windynation

PowerFlex Welding & Battery Cable Specifications

Part # BL = Black RD = Red	Size (AWG)	Stranding (0.25mm)	Insulation Thickness (inches)	Conductor Diameter (inches)	Approx. Total Diameter (inches)	Maximum Amperage
CBL-CSBL-08 CBL-CSRD-08	8	131	0.07	0.129	0.31	55
CBL-CSBL-06 CBL-CSRD-06	6	263	0.07	0.20	0.36	115
CBL-CSBL-04 CBL-CSRD-04	4	370	0.07	0.23	0.40	150
CBL-CSBL-02 CBL-CSRD-02	2	634	0.07	0.30	0.46	205
CBL-CSBL-10 CBL-CSRD-10	1/0	1004	0.08	0.37	0.56	285
CBL-CSBL-20 CBL-CSRD-20	2/0	1255	0.08	0.43	0.63	325
CBL-CSBL-40 CBL-CSRD-40	4/0	2047	0.08	0.56	0.75	440

Windynation welding cables current capacity

Temperature Correction

Another factor that needs consideration is the ambient temperature. Ambient temperature also influences the resulting temperature of the conductor, and it can increase it (hot climate) or decrease it (cold climate).

The electric current values of the previous table are calculated assuming an ambient temperature of 86°F (30°C). This may not be the temperature in your location or even inside the battery compartment. Therefore, a temperature correction factor must be applied. We can account for this using two methods: an equation or predetermined tables.

116

The equation method consists of applying the following formula:

$$I' = I \sqrt{\frac{T_c - T_a'}{T_c - T_a}}$$

Where:

I'= Ampacity corrected for ambient temperature.
I= Ampacity shown in the previous table.
Tc= Temperature rating of the conductor (°C).
Ta'= New ambient temperature (°C).
Ta= Ambient temperature used in the table (°C).

The I' value will represent the new ampere rating permissible for that conductor in that specific ambient temperature.

The formula looks intimidating, but it is easy to explain with an example.

Let us assume you need a copper THWN-2 #2AWG cable. At 194°F (90°C), it has an ampacity of 130A.

Let's assume the ambient temperature inside the battery compartment is 104°F (40°C). The expression would be as follows. The 30°C is from the standard NEC table.

$$I' = I \sqrt{\frac{T_c - T_a'}{T_c - T_a}} = 130\ A \times \sqrt{\frac{90°C - 40°C}{90°C - 30°C}} = 118\ A$$

This means the wire's new current-carrying capacity will be reduced from 130A to 118A. As you can see, the ambient temperature can have a significant impact.

The second method uses pre-established correction factors by temperature ranges using the values in the following table. This method can be easier.

For ambient temperatures other than 30°C (86°F), multiply the allowable ampacities specified in the ampacity tables by the appropriate correction factor shown below.				
Ambient Temperature (°C)	Temperature Rating of Conductor			Ambient Temperature (°F)
	60°C	75°C	90°C	
10 or less	1.29	1.20	1.15	50 or less
11–15	1.22	1.15	1.12	51–59
16–20	1.15	1.11	1.08	60–68
21–25	1.08	1.05	1.04	69–77
26–30	1.00	1.00	1.00	78–86
31–35	0.91	0.94	0.96	87–95
36–40	0.82	0.88	0.91	96–104
41–45	0.71	0.82	0.87	105–113
46–50	0.58	0.75	0.82	114–122
51–55	0.41	0.67	0.76	123–131
56–60	—	0.58	0.71	132–140
61–65	—	0.47	0.65	141–149
66–70	—	0.33	0.58	150–158
71–75	—	—	0.50	159–167
76–80	—	—	0.41	168–176
81–85	—	—	0.29	177–185

Ambient temperature correction factors for ampacity.
Source: National Electric Code

Visit this link for a better visual of this table:
https://cleversolarpower.com/temperature-correction

In this case, let's assume that the ambient temperature reduces to 59°F (15°C) and that your current demand estimation value is 140A, which makes you select a THWN 1/0 AWG cable rated for 75°C.

	Copper Conductors		
	Temperature Rating of Conductor		
SIZE	60°C	75°C	90°C
AWG OR kcmil	TYPES TW UF	TYPES RHW THHW THW XHHW THWN USE	TYPES RHH THHW RHW-2 THWN-2 XHHW THW-2 XHHW-2 THHN XHH USE-2
14**	20	20	25
12**	25	25	30
10**	30	35	40
8	40	50	55
6	55	65	75
4	70	85*	95*
3	85	100*	110*
2	95	115*	130*
1	110	130*	150*
1/0	125	150*	170*
2/0	145	175*	195*

Selecting a 1/0AWG cable

Since the manufacturer has rated the cable at 167°F (75°C) and the ambient temperature is 59°F (15°C), the temperature correction factor would be 1.15.

The next step is to apply the following expression:

$$Current = Current\ Load \times Correction\ Factor$$

$$150\ Amps \times 1.15 = 172.5\ Amps$$

We can see that this wire will be able to carry more current than it is rated for. The colder it is, the more efficiently it will work. We could use a 1AWG wire to carry 140 Amps of current in this situation.

$$130\ Amps \times 1.15 = 149.5\ Amps$$

However, when calculating this, you should always use the worst-case scenario. This means using the warmest value your wire will be susceptible to. Let's redo this calculation with realistic numbers.

Let's say the hottest time of summer is about 105-113°F (41-45°C). The table shows the corresponding temperature correction factor.

Ambient Temperature (°C)	Temperature Rating of Conductor			Ambient Temperature (°F)
	60°C	75°C	90°C	
10 or less	1.29	1.20	1.15	50 or less
11–15	1.22	1.15	1.12	51–59
16–20	1.15	1.11	1.08	60–68
21–25	1.08	1.05	1.04	69–77
26–30	1.00	1.00	1.00	78–86
31–35	0.91	0.94	0.96	87–95
36–40	0.82	0.88	0.91	96–104
41–45	0.71	0.82	0.87	105–113

Correlating temperature factor

$$150 \; Amps \times 0.82 = 123 \; Amps$$

We now see that this wire can only safely carry 123 Amps instead of 140 Amps at the specified temperature. We need to increase our wire sizes to account for this.

$$175 \; Amps \times 0.82 = 143.5 \; Amps$$

If we use a 2/0 AWG wire rated for 175 Amps, we will reach the required 140 Amps in the worst-case scenario. Therefore, you must use a 2/0 AWG wire rated at 75°C.

The NEC (National Electric Code) recommends applying a safety factor of 1.25 for simplicity. If your inverter draws 83.3A, you should multiply by 1.25 to determine the required cable capacity. This calculation results in 104A, so rather than selecting a cable rated for 83.3A, you'll need one that can handle 104A.

DC Voltage Drop

Another factor to consider when selecting wire gauge sizes is the voltage drop across the wire, which is referred to as losses in the cable. The voltage drop will increase if your cable diameter is small and long, like the cable from the solar panels to the charge controller.

For example, a single solar panel with an 18-volt output wired to the charge controller with a 5% voltage drop will lose 0.9 Volts.

$$18\,Volts \times \left(\frac{5}{100}\right) = 0.9\,Volts$$

The initial 18 Volts drops to 17.1 Volts, which is not ideal. To remedy this, we need to reduce the resistance in the wire by selecting a wire with a bigger diameter.

Voltage drops are associated with the wire size selected, the length of the cable, and the system's voltage. They are undesirable in a PV system because they lead to power losses. They also influence the minimum voltage input rating for various devices.

Therefore, you must calculate the voltage drop for a specific wire gauge and verify whether that voltage drop is permissible.

Common standards for this parameter establish that a voltage drop lower than 3% must be ensured between the modules and the charge controller.

To calculate the voltage drop, you must apply the following expression:

$$A = \frac{\rho \times 2 \times l \times I}{v \times Vsys}$$

Where:

A= Transversal section of the cable [mm^2]
ρ= Specific Resistance [$\Omega \cdot mm^2/m$]
$\qquad$ 0.0171 $\Omega \times mm^2/m$ for copper
$\qquad$ 0.026 $\Omega \times mm^2/m$ for aluminum
2= Total travel length for both + and - wire
l= Length of the cable [m]
I= Nominal current through the cable [A] (Isc*1.56)
v= Permissible voltage drop in the cable [no unit] (3% is 0.03)
Vsys= Open circuit voltage [V] (Vmp)

For example, we can consider two solar panels in series with 82ft (25-meter) cable length to the charge controller. The string has a short circuit current (Isc) of 5.8A and an open circuit voltage (Vmp) of 35V ($17.5V \times 2$).

The desired voltage drop in the system will be 3%.
Applying the expression:

$$A = \frac{\rho \times 2 \times l \times I}{v \times Vsys} =$$

$$\frac{0.0171 \left[\Omega \times \frac{mm^2}{m}\right] \times 2 \times 25m \times 5.8A}{0.03 \times 35V} = 4.7mm^2$$

As you can see, the result is given in mm². This represents the transversal section, where the copper wire must limit the voltage drop to 3%. We can convert the mm2 to AWG by referring to the following table. For this application, we would require a #10AWG size cable. Solar PV cables come in 12, 10, and 8AWG.

16	1.5mm²
14	2.5mm²
12	4.0mm²
10	6.0mm²
8	10mm²
6	16mm²
4	25mm²
2	35mm²
1	50mm²
1/0	55mm²
2/0	70mm²
3/0	95mm²
4/0	120mm²

AWG to mm² conversion table
Source: Cleversolarpower.com

The voltage drop is significant when the wire is long. This is from the solar panels to the charge controller. Or from the alternator to the battery. It doesn't need to be calculated for the rest of the system as long as the components are close together.

If you do not want to use the formulas, you can use this voltage drop calculator:
https://cleversolarpower.com/book/voltagedrop
We will also look at this tool in the next chapter.

Next, let's look at the wire types.

Wire Types

We can divide the type of wire that you need to use by sections of the PV system:

- Solar panels - Combiner box
- Combiner box - Charge controller - Loads
- Battery cables

Solar Panels – Combiner box

According to the UL-4703 standard, two types of wires must be considered for the connection of PV modules in PV applications.

The best type of wire for standard RV, boat, or roof applications is the PV wire. It is a single-conductor wire that connects the solar panels in series or parallel to the combiner box or directly to the charge controller.

These wires meet UL-4703 requirements and are made of either copper, aluminum, or copper-clad aluminum.

UL-4703 PV wire with MC4 connectors

Their insulation cover is based on cross-linked polyethylene (XLPE) or ethylene-propylene rubber (EPR), and they are rated to work at 600V, 1kV, or even 2kV.

The most important factor that makes these wires different from others is that they are designed to endure intense ultraviolet (UV) radiation to which the cables will be exposed.

Other wires would deteriorate over time and damage their insulation, which could cause exposure to the copper or aluminum material and lead to short-circuit failures. The temperature rating of these wires is close to 195°F (90°C) under wet conditions and between 220-300°F (105°C-150°C) under dry conditions. Also, a PV wire is designed to be flame-retardant, which is an essential feature for the safety of your system.

These wires must be used to connect the solar panels to the combiner box. If you do not have a combiner box, they can also be used to connect directly to the charge controller.

The second type of wire that can be used is the USE-2, which stands for underground service entrance wire.

The USE-2 wire is also used in solar PV applications, especially ground-mounted applications. The wires can go underground and are considered the alternative option when the PV wire (UL-4703) cannot be purchased. This type of wire is rated at 600V (suitable for solar purposes). They are cheaper than PV wires. However, they do not have UV protection. They are not flame-retardant; their maximum operating temperature under dry conditions is 195°F (90°C). USE-2 is also less flexible than PV wire.

It is recommended to use PV wire (UL-4703) because it is more temperature-resistant, flexible, and has thicker insulation. However, using the regular USE-2 wire is not forbidden.

Combiner box – Charge Controller – Loads

There will be no exposure to sunlight in this section, so wire insulation can be simpler. For mobile and boat applications, you should use marine-grade wires. These have smaller strands, which make them more flexible. The small strands will not break with vibrations.

Flexible marine-grade wire

You can use copper wires with a tin coating. This will make them more corrosion-resistant.

You can use a THW wire for cabins or tiny homes. These wires are rated at 75°C and are flame-retardant, moisture-resistant, and heat-resistant. They are also insulated with thermoplastic.

These wires are used for machine tools and the internal wiring of multiple appliances. They are rated to work at 600V, up to 195°F (90°C) in dry conditions, and 140°F (60°C) under wet conditions.

THW wire

These wires are less flexible than marine-grade wire because they have fewer but thicker wire strands. If you want to use very flexible wire, use marine-grade wire.

You can use these wires indoors. Do not use them outdoors or near your battery. There are other wires suitable for near batteries.

Battery Cables

Depending on your off-grid application, the batteries will generally be located in a separate compartment. During summer, these spaces could be exposed to high temperatures and an increase in humidity. Deep-cycle lead-acid batteries used for off-grid purposes occasionally expel internal gases into the environment. However, they would expose chemicals to the wires in case of damage or leakage.

Therefore, a THW is not suitable for this section. You must use a THWN-2 cable.

THWN-2 wire

THWN-2 stands for Thermoplastic Heat and Water-Resistant Nylon Coated. This cable is suitable for temperature ratings of up to 195°F (90°C) in dry and wet locations. They also have a flame retardant feature and a high resistance to abrasion from oil and chemical agents, thanks to the nylon coating.

These features make it ideal for applications where you need to connect batteries. Alternative options that can be considered are:

- THHN (Thermoplastic High-Heat Resistant Nylon Coated)
- XHHW-2 (Cross-Linked Polyethylene High-Heat and Water Resistance)
- Welding cables (rated up to 105°C or 221°F).

These wires must be used in the battery compartment between the charge controller and the battery system and for interconnections between batteries, whether you make series or parallel connections.

This means that the wire from the charge controller should be of this type. The wire that goes to the busbar, inverter, and DC fuse box should also be this type because they all start or end in the battery compartment.

As previously discussed, these wires use fewer strands because the individual wires are thicker. Use welding or marine-grade wires if you want very flexible wire or have an RV or boat.

Calculating Wire Sizes

In this chapter, we will calculate the wire size you need for each section of your solar system.

PV Modules – Combiner box

In this section, your reference must be the short-circuit current established in the datasheet of the PV module. Then, a security factor associated with higher irradiance levels and voltage drop must be applied. The maximum current a solar cell can produce is:

$$Imax = Isc \; x \; 1.56$$

SUNPOWER™

MODEL: SPR-E-Flex-100		
Rated Power (Pmax)[1] (+/–3%)	100	W
Voltage (Vmp)	17.5	V
Current (Imp)	5.80	A
Open-Circuit Voltage (Voc)	21.0	V
Short-Circuit Current (Isc)	6.20	A
Maximum Series Fuse	15	A

Standard Test Conditions: 1000 W/m², AM 1.5, 25° C
Suitable for ungrounded, positive, or negative grounded DC systems
Field Wiring: Cu wiring only, min. 12 AWG/4 mm², insulated for 90° C min.

⚠ WARNING ⚠
SEVERE ELECTRICAL HAZARD
• Solar module has full voltage even in very low light.
• Installation should only be done by a qualified technician.

524085
www.sunpower.com
Patented as shown at www.sunpower.com/patents

CE · RoHS

Sunpower 100-Watt PV module specifications

Calculate the maximum current through the wire:

$$Imax = 6.20A \times 1.56 = 9.7 Amps$$

For this section of the PV system, you will note that wire sizes with a manufacturer's temperature rating of 90°C can be as small as #14AWG.

However, this is without considering voltage drop. Therefore, we need to calculate the wire size to reduce the voltage drop to an acceptable 3% voltage drop.

$$\frac{0.0171 \left[\Omega \times \dfrac{mm^2}{m} \right] \times 2 \times 6m \times 9.7A}{0.03 \times 17.5V} = 3.79mm^2$$

3.79mm² is a #10AWG wire. This will only happen if you wire the cable for 20ft (6 meters) without an extension. Therefore, you must calculate this correctly to avoid unnecessary power loss. Get your voltage drop as low as possible while your wire is still affordable.

This wire will be used for every connection between modules and send the electric current to the fuses in the combiner box (if you have one) or straight to the charge controller.

I recommend using an online calculator for this. You can find it here: https://cleversolarpower.com/book/voltagedrop

Let's repeat the same calculation using this calculator.

First, you enter all your details: the distance to the charge controller, voltage (Vmp), and current (Isc*1.56). Select DC current as well. Then you press calculate, and you will see the voltage drop percentage change. Keep lowering your wire diameter size until this value goes below 3%. You will then find your required wire size.

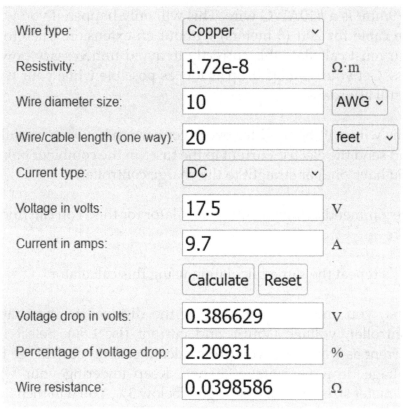

Wire type:	Copper ⌄	
Resistivity:	1.72e-8	Ω·m
Wire diameter size:	10	AWG ⌄
Wire/cable length (one way):	20	feet ⌄
Current type:	DC ⌄	
Voltage in volts:	17.5	V
Current in amps:	9.7	A
	Calculate Reset	
Voltage drop in volts:	0.386629	V
Percentage of voltage drop:	2.20931	%
Wire resistance:	0.0398586	Ω

Voltage drop calculator
Source: www.rapidtables.com

- Wire type: usually copper.
- Resistivity: standard copper resistance (do not change).
- Wire diameter: this is the value you will adjust after looking at the 'percentage of voltage drop'.
- Wire length: one-way wire length from solar panels to the charge controller.
- Current type: select DC current.
- Voltage: the Vmp of the panel(s).
- Current: this is the Isc of your panel x 1.56.

You will then press calculate and adjust the 'wire diameter number' to reduce the 'percentage of voltage drop' to under 3%.

Combiner box – Charge Controller

The output of the combiner box will go to the charge controller. This section will contain all the electric current that flows from the solar panels. You do not need to use this formula if you wire your panels in series. Instead, to estimate the wire size, we must use the following formula:

$$Ibox = Isc \times number\ of\ strings \times 1.56$$

The number of strings represents the number of parallel connections made in the solar array, and the factor 1.56 is associated with safety factors according to the extra heat these wires have to withstand.

For example, one array of two panels in parallel:

$$Ibox = 6.2\ Amps \times 2\ strings \times 1.56 = 19.37 Amps$$

Calculate the voltage drop if the wires travel 16ft (5 meters) from the combiner box to the charge controller:

$$\frac{0.0171 \left[\Omega \times \frac{mm^2}{m}\right] \times 2 \times 5m \times 19.37A}{0.03 \times 17.5V} = 6.3mm^2\ or\ \#8AWG$$

The voltage will increase if your panels are connected in series, but the current will stay the same. **That is why a series connection is always preferred to limit the voltage drop.**

A solar panel series connection requires a wire with a smaller diameter. This is because the wire diameter is decided by the current that runs through the wire. In parallel, amps will be added up while the voltage stays the same. In series, the voltage will be added up while the current remains the same. More on this later in the book.

Charge Controller – Battery

The current that flows from your charge controller to the batteries will have a maximum charging current listed in its datasheet.

Let's say you have a 40A charge controller. The charge controller cannot deliver more than 40A to the battery. Therefore, we must apply the standard safety factor of 1.25 to size this cable.

$$40A \times 1.25 = 50A$$

We will select a 50A fuse and a cable that can carry at least 50A (more on this later).

Battery – Inverter

To size the wires that go from your batteries to the inverter, you need to know the power rating of the inverter you will use. For example, if you are using a 1,500 Watt inverter with a 12 Volt battery bank, you apply the following formula:

$$Current = \frac{Power}{Voltage}$$

$$\frac{1,500\ Watts}{12\ Volts} = 125\ Amps$$

Then you have to use the 1.25 safety factor to size the correct cable:

$$125A \times 1.25 = 156A$$

First, we select a fuse close to 156A (but higher than 156A). If we are going to use MEGA fuses, we can choose from the following:

40A, 60A, 80A, 100A, 125A, 150A, **175A**, 200A, 225A, 250A, 300A, 350A, 400A, 450A, and 500A

We will select a 175A MEGA fuse. A cable that can carry at least 175A is a 2AWG welding cable from Windynation.

windynation

PowerFlex Welding & Battery Cable Specifications

Part # BL = Black RD = Red	Size (AWG)	Stranding (0.25mm)	Insulation Thickness (inches)	Conductor Diameter (inches)	Approx. Total Diameter (inches)	Maximum Amperage
CBL-CSBL-08 CBL-CSRD-08	8	131	0.07	0.129	0.31	55
CBL-CSBL-06 CBL-CSRD-06	6	263	0.07	0.20	0.36	115
CBL-CSBL-04 CBL-CSRD-04	4	370	0.07	0.23	0.40	150
CBL-CSBL-02 CBL-CSRD-02	2	634	0.07	0.30	0.46	205

Windynation cable selection

One advantage of using a system with a higher voltage is that you don't need to use big wires. If you have the same inverter with a battery bank of 24Volts, you only need a wire that is capable of carrying 78A Amps, as can be seen in the following calculation:

$$\frac{1,500\ Watts}{24\ Volts} = 62.5\ Amps\ \times 1.25 = 78A$$

Then, you would have to use a 100A MEGA fuse with a 6AWG welding cable.

It is not recommended to use a cable carrying more than 100 Amps in a DIY system. If your wire carries more than 100 Amps, use a system with a higher voltage. This will not only increase safety but also save you on wire costs.

Interconnecting Batteries

The cable size must be the same when you connect batteries in series.

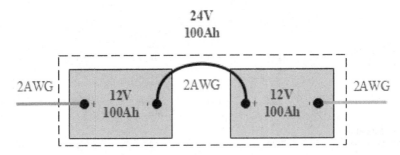

Two 12V batteries in series require the same cable thickness

We can see that these two 12V batteries make a 24V battery. Since they are in series, the current will pass through both batteries, and the current will be the same.

But if you connect in parallel, you can make the wires smaller. Because each battery delivers a share of the current in parallel. This is only true when you wire to a busbar. Let me show you the difference.

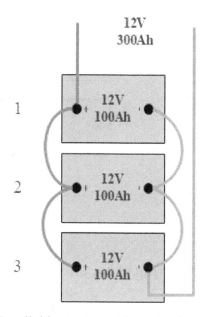

Parallel batteries with no busbar

The current going through the wires will be the same because the current from battery 3 has to go through the connections of batteries 2 and 1. So if battery 3 delivers 33% of the current, it has to pass through the cables of batteries 2 and 3 to make 100%.

On the other hand, if you use a busbar, the current in the wires will be reduced. As you will soon learn, I recommend fusing every parallel battery. It's best to use a fused busbar for this. The wires coming from the battery will only carry 33% of the total current.

Take a look at the following diagram:

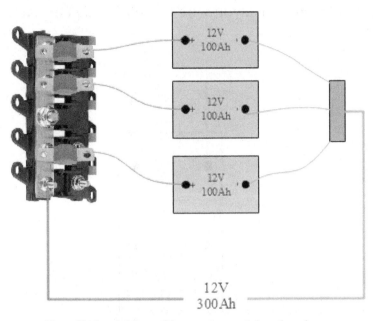

Parallel wiring of batteries with a busbar

If the inverter draws 100A, every battery delivers 33% of the current. However, because batteries can have slightly different internal resistance, I advise using a safety factor of 1.56 instead of the usual 1.25.

So if we have a 1,000W inverter, the current is:

$$\frac{1,000W}{12V} = 83A$$

We have three batteries in parallel, so every battery is expected to deliver 33%.

$$\frac{83A}{3 \ batteries \ in \ parallel} = 28A \ per \ battery$$

Then, we add the safety factor of 1.56:

$$28A \times 1.56 = 44A$$

We then have to size a fuse. The nearest fuse (rounded up) is 50A. A cable that can carry at least 50A is an 8AWG or 10mm² welding cable.

windynation

PowerFlex Welding & Battery Cable Specifications

Part # BL = Black RD = Red	Size (AWG)	Stranding (0.25mm)	Insulation Thickness (inches)	Conductor Diameter (inches)	Approx. Total Diameter (inches)	Maximum Amperage
CBL-CSBL-08 CBL-CSRD-08	8	131	0.07	0.129	0.31	55

Selection of an 8AWG welding cable

If one battery shuts down, the other two will share the current at 50% (41.5A). It could be that the fuse blows. But that is what you want; otherwise, you will not be notified of the fault.

Wiring for Electrical Loads

This section applies to DC and AC loads.

Typical DC loads will work with large sizes like #8AWG or even #2/0AWG wire sizes.

Typical AC loads, such as lighting, TVs, microwaves, fans, small motors, and others, will generally use #14AWG or #16AWG wire sizes, typically used for outlets.

Depending on the model, A/C units, washing machines, and refrigerators may require #12AWG.

You can find the electric current demand for your appliance on the product or in the product's datasheet. If you are only presented with the power rating, use the following formula to figure out the current:

$$Current = \frac{Wattage}{Volts}$$

For an AC appliance:

$$\frac{1,000\ Watts}{120\ Volts} = 8.33\ Amps$$

For a DC appliance:

$$\frac{1,000\ Watts}{12\ Volts} = 83.3\ Amps$$

Wire Safety

Having a lot of wires close to each other will be inevitable in compact off-grid solar installations. Therefore, there are some rules you need to follow, especially if your system has vibrations like an RV or a boat.

1. Always protect your wires with a wire sleeve when they pass a bend or another wire. This prevents the insulation from being cut, which can cause a short.
2. Do not make sharp bends with wires. Over time, this can damage the wire strands and the insulation.

3. If possible, run your wires through a conduit. It's easier to add more wires later on.

PVC and braided wire sleeve

Batteries

How do Batteries Work?

Batteries work through an electrochemical process that involves a double conversion of energy. The first conversion is to charge the battery and change from electrical to chemical energy. The second conversion process is developed from chemical energy to electrical energy. This is done during the discharge process.

All batteries operate on this principle. To make this energy conversion possible, two electrodes from different metal components must act as the positive and negative terminals of a voltage source. Also, an ionic medium must connect both electrodes. This is commonly known as the electrolyte, a liquid composition that allows the transfer of electrons between electrodes. This whole arrangement receives the name of the *voltaic cell*. The collection of several cells makes the battery.

A direct current voltage source must be connected with the correct polarity to the terminals (electrodes) to fully charge the battery. The voltage of the DC source must be higher than the battery's voltage.

Lead-Acid

As stated before, lead-acid batteries have two electrodes. One electrode acts as the positive terminal and is filled with lead oxide (PbO2), and the other acts as the negative terminal and is made of pure lead. That's why lead-acid batteries are heavy.

The medium used for lead-acid batteries is sulfuric acid. Lead-acid batteries can be classified into two main groups:

- Flooded (VLA or valve lead-acid)
- Sealed (VRLA or valve-regulated lead-acid)

Let's look at them in more detail.

Flooded

The first type of lead-acid battery is the flooded or valve lead-acid battery (VLA). VLA batteries are identified by having small, ventilated access to their internal structure with removable plugs that allow verification of the battery's specific gravity and state of charge.

The main downside of these batteries is that they emit gases generated by internal electrochemical reactions. Due to this, these batteries must be located in an area with vented access where air circulates constantly. The accumulation of these gases within a small, closed area can be dangerous.

Flooded deep-cycle lead-acid battery

Internal electrochemical reactions and gas expulsion also reduce water levels inside the battery. This is why periodic maintenance involves adding distilled water to the battery through the caps on top.

Maintenance also involves cleaning the terminals to remove oxide that has accumulated over time. You can recognize flooded batteries if you see removable caps on top. The caps can be removed easily with a twisting mechanism. Do not confuse them with sealed lead-acid batteries, where the caps are not meant to be removed.

Keep in mind that VLA batteries must always be placed upright and vertically. Otherwise, the internal fluid will spill out. At the same time, vented batteries can also be classified into three categories.

- **Ignition Batteries**

These batteries are used for automotive purposes. They seem to be economical for solar builds, but they are not.

Ignition batteries are made to deliver a high amount of current for a short time. This makes their construction different and makes them unsuitable for long-duration low-current applications. They are also called SLI batteries.

- ## Deep Cycle Batteries

Deep-cycle batteries are used for applications where low amounts of current are needed but for longer periods. These batteries can endure many charge and discharge cycles. Since these are requirements used in photovoltaic applications, this lead-acid battery is used for solar power storage.

- ## Backup Batteries

These batteries provide energy for control operations and backup in substations. They are generally not used for off-grid solar power applications.

You might be able to pick these up when a power company, factory, or hospital wants to switch its backup battery bank every few years. However, they are not considered the first choice for lead-acid batteries in solar systems.

- **Sealed (VRLA)**

Sealed is the second type of lead-acid battery. These are partially sealed to avoid the evaporation of the electrolyte. VRLA stands for valve-regulated lead acid. They have a pressure-sensitive valve that automatically controls the emission of gases, but in normal operation conditions, they are closed. They are opened automatically to release gases in case there is high pressure inside the battery. This will happen if something is wrong with the battery, like a short circuit.

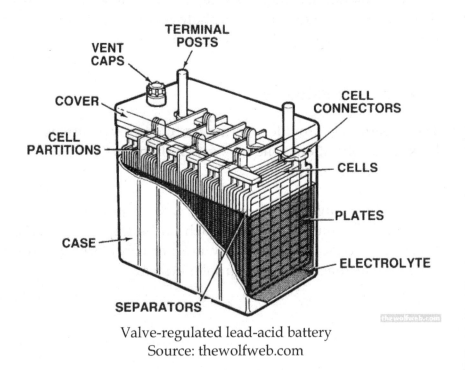

Valve-regulated lead-acid battery
Source: thewolfweb.com

These batteries recombine oxygen and hydrogen through an electrochemical process that allows them to reincorporate water back into the cell instead of evaporating and constantly releasing it, as the VLA batteries do.

146

Thanks to the pressure valve, gas emissions and contaminations are minimal. The downside of these batteries is that, in some cases, they are more sensitive to the operating temperature than flooded batteries.

Depending on the electrolyte, there are two other types of VRLA batteries: Gel and AGM.

- **Gel Cell**

These batteries have a silicon compound. When added to the liquid electrolyte, the substance acquires a gel-type consistency. Gel cell batteries have a longer lifespan than flooded-type batteries and guarantee more charging and discharging cycles. This gel consistency is also a significant advantage. Even when the battery is not placed upright or if the battery case breaks or deteriorates, the sulfuric acid will not spill because the gel structure is solid.

Gel-type batteries also withstand deep cycle discharges with high temperatures and even vibrations. Besides, they have a stable voltage during the discharge and do not require maintenance, as vented batteries do.

They are also corrosion-resistant and resistant to lower temperatures. However, their main disadvantage is high internal resistance, which translates into a lower discharge current (C-rate). Gel batteries have a life cycle of around 1,000 cycles when discharged to 50%.

- **Absorbed Glass Mat (AGM)**

In AGM-sealed batteries, the electrolyte is absorbed by fiberglass, which works as a sponge that immobilizes sulfuric acid. AGM batteries offer the same benefits as gel cell batteries. The difference is that AGM batteries withstand higher charging voltages than gel types but are not immune to high voltages. They will still offgas.

Due to their structure, AGM batteries have a lower internal resistance, which is why they can deliver or absorb higher electrical currents during charging and discharging than gel-type batteries. A lower internal resistance also increases efficiency.

Unlike VLA batteries, these models can be placed vertically or horizontally. The charging current represents 20% of the battery capacity in Ah (0.2C). The AGM battery is the best choice if you use lead acid. When discharged to 50%, it has a cycle life of 1,000 cycles.

Lithium Batteries

Lithium batteries are the ultimate technology used in solar power applications. There are many lithium chemistry configurations, but Lithium Iron Phosphate (LiFePO4) is the predominant technology.

Lithium batteries have the highest depth of discharge (about 80-100%). They also have low self-discharging rates (they can be stored for several years) and a high energy density, which means smaller dimensions and less weight.

They have a superior efficiency (95-99%) and a low internal resistance, allowing them to charge and discharge at higher currents than lead-acid (1C).

Another essential advantage of lithium batteries is that they can deliver more charging and discharging cycles than lead-acid batteries.

An important fact for RV applications is that lithium batteries can weigh less than half of an equivalent lead-acid battery.

It used to be that lithium was many times more expensive than lead-acid, but at the time of writing (November 2024) a renogy AGM (lead-acid) 12V 100Ah battery sells for $190, while a reputable 12V 100Ah LiFePO4 battery sells for $200. Only a $10 difference, and you can use 100% of a lithium battery instead of a lead-acid's 50%.

LiFePO4 12V battery from Victron Energy

Lithium batteries are the most powerful, efficient, and long-lasting solution for solar power applications, including RV applications.

To make a higher-voltage system, like 24 or 48 volts, ensure the lithium battery can be connected in series. Some lithium batteries cannot be connected in series because of the battery management system. The battery's datasheet can verify this. However, if you are making a 24 or 48V system, I recommend using a 24V or 48V battery and not wiring 12V batteries in series because you will require a balancer if you do.

If you purchase a lithium battery, ensure it has a built-in low-temperature disconnect sensor (in the battery management system). If you charge a lithium battery in freezing temperatures, the battery will be irreversibly damaged. Therefore, watching the battery temperature and seeing if it has a built-in low-temperature cutoff is essential.

Some charge controllers come with a temperature sensor. This can tell the controller not to charge the battery when it's below or close to freezing. Of course, discharging below freezing is still possible.

You can also insulate the battery compartment and place some heat mats with a built-in thermostat to keep the battery above freezing. You can find these heat mats in your gardening store. They are typically used for germinating seeds.

Lithium is becoming the preferred choice for energy storage in off-grid systems. This is partly because of the rise in DIY lithium batteries. You can create your own 12V battery by wiring four 3.2V cells in series. You also need to add a BMS (battery management system), which keeps the voltage of every cell the same. These systems beat regular lead-acid batteries.

At the time of writing (November 2024), four 3.2V 280Ah cells cost around $200. Adding a 100Amp BMS to it will cost another $100, making it an attractive solution for power storage.

4x 3.2V LiFePo4 cells and a Daly BMS

There is an extensive online community about DIY LiFePO4 batteries. However, for safety reasons, not everybody wants or can create their own battery. You can buy a lithium battery online if you don't like to do it yourself.

I made a video on my YouTube channel comparing DIY vs. BUY batteries, and it became attractive for batteries over 5kWh.

There are a few recommended brands for buying raw LiFePO4 cells from this book's end. Let's continue with a comparison between lead-acid and lithium.

Lead-Acid vs. Lithium

This chapter was originally written in 2020, when I compared both. Lead-acid was still cheaper then per usable watt-hours, but now (November 2024), lithium is the clear winner, as you will see in the updated comparison.

Let's explore this by comparing the two.

Lead-acid discharge rate: 50%.
Lithium discharge rate: 80%.

Lead-acid weight: Heavy.
Lithium weight: Half the weight of lead-acid.

Lead-acid cycles: 1,000 if you discharge it to 50%.
Lithium cycles: 8,000 if you discharge it from 10-90%. If you cycle from 0-100%, it has 4,000 cycles. After the mentioned cycles, it still has 80% of its capacity left.

Lead-acid maintenance: Self-discharges over time.
Lithium maintenance: Can be stored longer without charging.

Lead-acid venting: Needs to be able to vent gases (depending on model).
Lithium venting: doesn't vent gases.

Lead-acid efficiency: 80-85%
Lithium efficiency: 95 – 99%

Let's compare a renogy AGM with a reputable brand lithium battery.

Renogy AGM: 12V, 100Ah for $189 at 63.9 lb. / 29 kg
Litime LiFePO4= 12V, 100Ah for $199 at 26 lbs. / 11.8 kg

We can see that AGM is cheaper than Lithium. But we need to do some more calculations to figure out the true cost of the battery. First, let's address the fact that you can only use 50% of the AGM battery, while lithium can use 80% of its capacity. This means that we need about two AGM batteries for the same Lithium. The weight of your battery bank will now be 128 lb compared to 26 lb for lithium.

If we do a quick calculation, we will need almost two AGM batteries for one lithium. This will cost us $468 for AGM and $1000 for Lithium. These calculations refer to the actual cost per usable Ah.

Lead-acid: $189 $\times$ 2 (50% DOD) = $378

Lithium: $199 $\times$ 1.25 (80% DOD) = $249

We can already see that a lithium battery is cheaper than a lead-acid battery. But let's continue the calculation.

Next, we need to factor in the battery's efficiency. With lead-acid AGM, this is about 85%, and lithium has 99% efficiency.

Lead-acid: $378 $\times$ 1.18 (85% eff) = $446

Lithium: $249 $\times$ 1.01 (99% eff) = $252

Then we can address the most significant factor for AGM compared to Lithium. This is the number of cycles the battery can deliver.

For an AGM with a DOD (depth of discharge) of 50%, this is a maximum of 1,000 cycles. You need to dispose the battery afterward. For lithium with a DOD of 80%, this is 8,000 cycles. But after those 8,000 cycles, the battery still has 80% of its rated capacity left, so you can still use it.

Lead-acid: $446 $\times$ 8 (1,000 $cycles$) = $3,568

Lithium: $252 $\times$ 1 (8,000 $cycles$) = $252

The cost of AGM is $3,568, and the cost of lithium is $252. Lead acid is **14 times** more expensive than lithium (LiFePO4).

Keep in mind that 8,000 cycles are about 22 years if you cycle the battery from 10-90% every day. In practice, the battery BMS will be defective before the cells do. So, a lithium battery's lifespan depends on the BMS's longevity. I expect a lifetime of around 10 years (the same lifespan as a solar string inverter).

Let's see how much the price difference is between the lead-)acid battery and 10 years of life expectancy.

Lead-acid: $446 × 3.65 (1,000 *cycles*) = $1,628

Lithium: $252 × 1 (3650 *cycles or* 10 *years*) = $252

We can see that lithium is still the clear winner even if the battery fails after 10 years.

Building your own DIY Lithium battery can drive this cost down even more.

Temperature and Lead-Acid Batteries

Lead-acid batteries are highly temperature-dependent. Each battery is designed with a specific capacity in ampere-hours (Ah), but this capacity can be drastically affected by the internal operating temperature.

Temperature profoundly influences chemical reactions. Since lead-acid batteries' charging and discharging are based on chemical reactions, temperature plays an important role.

High temperatures result in an enhanced reaction rate, which increases the battery's instantaneous capacity. Conversely, it can also drastically reduce the life cycle.

Every 10°C increase concerning the optimum operating temperature value of 77°F (25°C) reduces the life of a battery to half of its rated lifespan. In other words, if an AGM battery is supposed to last nearly 10 years operating at 77°F (25°C), then increasing the operational temperature value to 95°F (35°C) will reduce its life expectancy to 5 years. As you can see, the effects can be devastating for life expectancy.

On the other hand, low temperatures also impact battery capacity. The nominal reference at 77°F (25°C) equals 100% capacity. As can be seen in the following figure, increasing the temperature also increases the battery's available capacity (up to 120% at 122°F or 50°C).

However, as temperature reduces, the percentage of nominal capacity that the battery will have available will reduce. Capacity reductions could reach 65% of the nominal capacity at -4°F (-20°C).

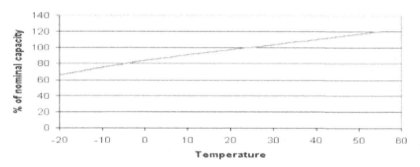

Effect of temperature (°C) on effective capacity
Source: Sandia Report

Apart from capacity, the temperature can also affect the efficiency values, as seen in the following figure.

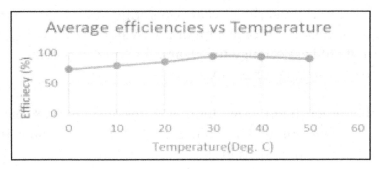

Average efficiencies vs. temperatures.
Source: allsciencejournal.com

If ambient temperature values cannot be modified, then the way to compensate for internal temperatures is to use lower or higher charging currents.

Under low temperatures, the batteries should be charged at higher voltages and higher current values. Meanwhile, if high temperatures are present, the charging current should be reduced to 75% of the rated current.

This is something that the charge controller will do. The device must have a temperature sensor with a temperature compensation feature (more on this later).

For lead-acid batteries, it is recommended that the ambient temperature be between 77-86°F (25-30°C). Therefore, it's better to keep these batteries at room temperature instead of in a box outside.

When the winter arrives, you will likely need to store your RV and battery bank as well. When doing this, remember that the battery must be fully charged before storing it.

During storage, you must keep the battery in a dry and warm indoor location. The stored battery will self-discharge over time. You must use a smart charger that keeps the state of charge full during storage. A smart charger (or trickle charger) will periodically charge your battery without leading to overcharging.

Lithium batteries have a higher tolerance for temperatures. The best operating temperature is between 59-95°F (15-35°C). They can operate in a wider temperature range of 32-95°F (0-35°C) and can be stored at temperatures between -4 -131°F (-20-55°C). Lithium batteries cannot be charged if they are below freezing, as this will destroy them.

Another aspect of lead-acid batteries is that it is harder to charge a battery from 90% to 100% than to charge a lithium battery from 90% to 100%. This is because the internal resistance of the lead-acid battery is high when it is near full capacity, whereas with lithium, the internal resistance is low. Therefore, charging from 90% to 100% would be the same as charging from 50% to 60%.

Think of lead-acid batteries as a football stadium that needs to be filled. When the stadium is almost full at 90%, the last 10% of the people need to step over other people's feet to get to their seats. This will take longer than the first 10% of people who entered the stadium.

If we compare lithium to a football stadium, people would enter the stadium with no empty seats. So, there is no need to step over other people's feet. This way, the stadium will fill more quickly without stepping over other people's feet.

Never put lead-acid batteries on a concrete floor. The cold from the concrete can damage the batteries on the bottom, reducing their capacity. Use a wooden board to insulate the bottom of the batteries.

Using car Batteries

Some think a car battery can power appliances in a small off-grid PV system. However, this is a bad choice.

Car batteries are ignition batteries, which were discussed previously in the vented battery type section.

Ignition batteries are characterized by delivering a large amount of current needed to start the engine instantly. In other words, car batteries are designed to provide large amounts of current for a very short time. The plates in a starter battery are thicker to accommodate the larger currents. A deep-cycle lead-acid battery has thinner plates suited for long-term lower current delivery.

Batteries needed for solar power applications do not work the same way. To keep the system running, lower amounts of current and longer periods of use are required. Batteries must also have a deep discharge rating and higher conversion efficiencies.

Therefore, choosing a car battery will have low reliability in the long, medium, and short term for solar power applications.

Series and Parallel

Batteries can be connected in three configurations:

- Series
- Parallel
- Combination of series and parallel

Like solar panel connections, wiring your batteries in series will increase the battery bank's voltage while keeping the battery's energy capacity (Ah) the same. To connect your batteries in series, you must wire the positive terminal with the negative terminal of the other battery. The following image shows a schematic of the series connection.

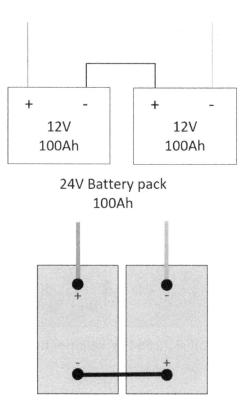

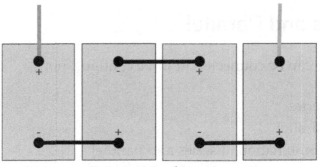

Batteries joined in series

I recommend avoiding connecting 12V batteries in series to create a higher voltage, as this setup often leads to imbalance issues between the batteries. When two 12V batteries are connected in series to create a 24V system, they're seen as a single 24V battery by the charge controller or charger. However, differences in internal resistance between the two batteries can lead to one battery reaching a higher voltage, such as 13.3V, while the other lags behind, perhaps at 13.0V. This disparity is what we call an imbalance.

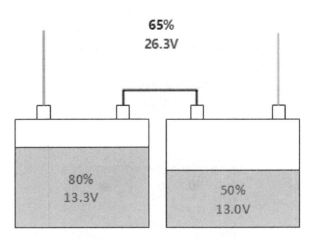

Two unbalanced 12V batteries in series

For instance, if battery 1 reaches 100% charge at 13.6V, battery 2 might only reach around 70% charge at a lower voltage. As a result, battery 2 never fully charges to 100%, so you're missing out on a significant portion of its capacity.

While a battery balancer can help by transferring energy from the more charged battery to the less charged one, adding this component introduces extra complexity. Properly designed systems, such as a single 24V battery with an integrated BMS (Battery Management System), won't need a balancer, simplifying the setup.

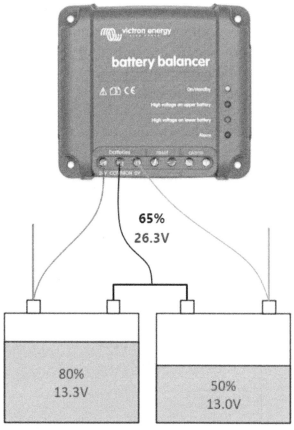

Adding a balancer to 12V batteries in series

For those who use a balancer, it's important to note that it activates around 27.3V and deactivates at 26.6V (in the case of Victron). This means the balancer only engages when the batteries actively charge through solar, shore, or generator power.

Additionally, if you still decide to use two 12V batteries in series, fully charge them before connecting. Starting with partially charged batteries in series will lead to an initial imbalance that can compromise performance over time.

On the other hand, a parallel connection consists of joining the positive terminal of one battery with the positive terminal of the second battery and the negative terminal of the first battery with the equivalent negative terminal of the second one. This connection maintains the battery voltage but increases the battery bank's energy capacity (Ah).

It is essential to use both terminals of the outer batteries as the main terminals instead of the terminals of one battery.

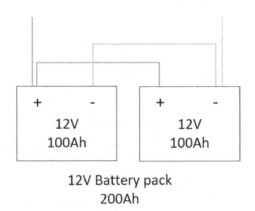

12V Battery pack
200Ah

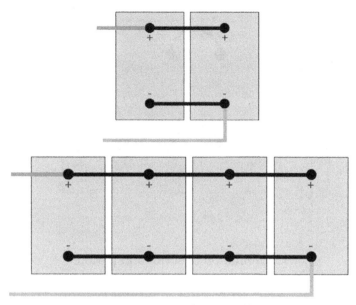

Batteries joined in parallel

Finally, the series/parallel connection combines the two when both voltage and energy capacity need to be increased. This consists of making the series connections first and then joining the endpoints of the string of batteries according to their polarity.

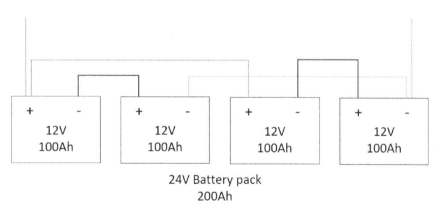

24V Battery pack
200Ah

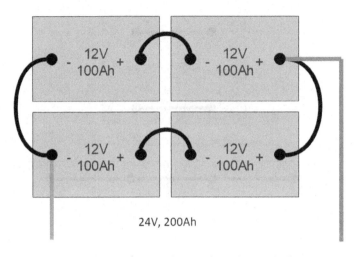

24V, 200Ah

Combination of series and parallel

Connecting batteries with different voltages should not be done in series or parallel. Mixing voltages causes imbalances and uneven charging/discharging. Higher-voltage batteries will push current into lower-voltage ones, damaging both batteries, reducing efficiency, and leading to safety issues like overheating or battery failure.

Connecting batteries with different capacities should not be done in a series connection. Because all batteries discharge at the same current, meaning the lower-capacity battery will reach its limit sooner, which can over-discharge and damage it.

However, you can connect batteries with different capacities in parallel. The smaller battery (on the next image) will deliver 33% of the current, while the larger battery (on the right) will deliver 66% of the current. However, because the internal resistance of the batteries is very low, the current from each battery depends mainly on the resistance of the cabling.

I have made a video about different capacity batteries in parallel, where I go much more in-depth on calculations. You can view it here:
https://cleversolarpower.com/book/differentcapacity

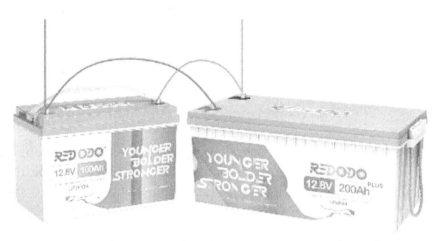

One 12V 100Ah in parallel with one 12V 200Ah Lithium battery

Never use one battery as the main battery terminal in a parallel configuration. In the examples before, the positive wire is on the first battery, while the negative wire is on the last battery. This is because of the resistance of the connectors and wires.

If you wire four batteries in parallel and use the left battery as the main terminal, the left battery will work much harder than the battery on the right. This will lead to imbalances in the batteries and early death for the left battery.

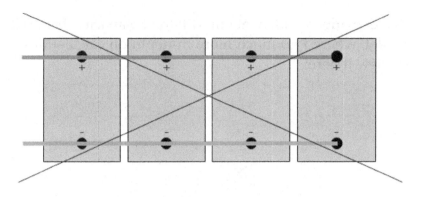

Never connect multiple batteries like this

Make sure that the interconnecting cables are the same length and size. When they are the same, the resistance will be the same, which is beneficial for the whole battery because the resistance will be the same everywhere.

C-rate

We have discussed the battery's discharge rate before but have not described it yet. It is time to do so.

As explained before, batteries depend on a chemical reaction to generate electricity. Therefore, a battery's available capacity depends on how quickly it is charged or discharged relative to its total nominal capacity (Ah).

In other words, if you discharge a battery very quickly, its capacity will be lower than the indicated capacity. This is because of the chemical reaction that generates heat inside the battery. Also, the higher the discharge rate, the lower the battery's voltage due to a voltage drop at the terminals. Potentially turning off your inverter when a large load is placed on a small battery.

The battery's total capacity can be briefly abbreviated as C and represents how much energy can be stored inside the battery. For example, a battery with a capacity of 100Ah is the same as C=100.

The charge and discharge rates of a battery are measured in C-rates. The manufacturer always provides the C-rates, and each battery has a nominal current value for each C-rate.

There are different C ratings for batteries. Some say C_{20}, and some say $_{10}C$. What is the difference between having the number before or after the letter 'C'?

If the C-rating is in front of the letter C, you multiply it with the battery capacity. If the number is to the right of the letter 'C,' you divide it with the battery capacity.

If the battery has a capacity of 65C, you multiply the capacity by 65. This high C-rate is only found in small Lipo batteries for drones.

1800 mAh 65C lipo battery for drones

Let's calculate the discharge rate for this small battery pack.

$$1.8Ah \times 65 = 117\ Amps$$

This battery can provide 117 Amps continuously until it's empty. After that, however, the battery will get hot, and in practice, the battery capacity will be reduced because of the fast discharging.

One of the most common C rates manufacturers mention for off-grid lead-acid batteries is the C_5 rate, or 0.2C.

If you have a 100Ah battery with a rating of C_5, it can theoretically supply 20 Amps for 5 hours.

$$\frac{100\ Ah}{5} = 20\ Amps\ for\ 5\ hours$$

Why is the C rating important?
If you purchase a lead-acid battery, the capacity will be tested with a 20-hour discharge time (C_{20}) at a temperature of 68°F (20°C). Discharging a medium-sized battery over a period of 20 hours will not create much heat and will work close to its best efficiency.

If you take that same battery and discharge it over 5 hours instead of 20, its capacity will be reduced. This is because the battery must provide energy faster, which will increase the current, heat, and internal resistance of the battery.

The next image shows a 420Ah battery. As you can see, this battery is rated at 20 hours or C20. If we discharge the battery in 5 hours, we only get 344Ah. If we discharge the battery in 100 hours, we get 467Ah, which is 47Ah more than the nominal capacity at the 20-hour rate.

Capacity ⁸ Amp-Hours (AH)			
5-Hr	10-Hr	20-Hr	100-Hr
344	386	420	467

Battery capacity relating to discharge time 6V 420Ah battery
Source: trojanbattery.com

Let's take this battery (6V), and see how much energy we can get from it at different C-rates.

C-rating	Capacity	Current
C_5	344 Ah	68.8 Amps
C_{10}	386 Ah	38.6 Amps
C_{20}	420 Ah	21 Amps
C_{100}	467 Ah	4.67 Amps

We can see that the C_{20} rating is used to describe the battery's capacity, which is 420Ah. Let's see what will happen if we discharge it at a higher C rating.

The 6 Volt battery will drain from 6.36 Volts (100%) to 6.05 Volts (50%) over time. Draining a battery to 0% is not recommended (for lead-acid) because the battery will get damaged internally. We can use this to calculate the total energy stored inside the battery.

$$Energy = Voltage \times Current \times Hours$$

$$6.05V \times 68.8A \times 5\ hours = 2,081\ Watt\ hours$$

$$6.05V \times 38.6A \times 10\ hours = 2,335\ Watt\ hours$$

$$6.05V \times 21A \times 20\ hours = 2,541\ Watt\ hours$$

$$6.05V \times 4.67A \times 100\ hours = 2,825\ Watt\ hours$$

As you can see, longer discharge times are equivalent to higher effective energy capacity storage but imply lower instantaneous electric current demands.

Meanwhile, shorter discharge times allow for a higher instantaneous electric current demand, but they will reduce the effective capacity of the battery. The reduction effect is non-linear.

You will also see mentions of 1C or 0.5C, which is common in lithium batteries. For example, a lithium battery with a capacity of 100Ah, which has a discharge rate of 1C, we will be able to draw:

$$100Ah \times 1C = 100\ Amps$$

A battery with a discharge rating of 0.5C will be able to draw:

$$100Ah \times 0.5 = 50\ Amps$$

It is essential to look at your battery and the recommended C-rate. For lead-acid, this is usually C5 or C20 and for lithium 1C. This table will solve some confusion about C-rates:

C-rate		Capacity	Formula	Amps	Time
20C		100 Ah	20 x 100Ah	2000A	3 minutes
10C		100 Ah	10 x 100Ah	1000A	6 minutes
5C		100 Ah	5 x 100Ah	500A	12 minutes
1C		100 Ah	1 x 100Ah	100A	1 hour
0,5C	C/2	100 Ah	100Ah / 2	50A	2 hours
0,2C	C/5	100 Ah	100Ah / 5	20A	5 hours
0,1C	C/10	100 Ah	100Ah / 10	10A	10 hours
0,05C	C/20	100 Ah	100Ah / 20	5A	20 hours

C-rates in batteries

C-rates are used for discharging and charging batteries. Refer to the datasheet of your battery to see the recommended charging rate. It is important to check that the current going into the battery is not too high.

Lithium batteries have a high C-rate for charging and discharging because they have low internal resistance. On the other hand, lead-acid batteries will have a lower C-rate for charging and discharging because the battery's internal resistance is higher. Another good reason to choose lithium (LiFePO4 over lead-acid).

Battery Monitors

One of the most useful gadgets for solar power is a battery monitor. This is usually a feature of a charge controller, which includes a battery voltage that provides the user with a reference to the battery bank's state of charge.

Each lead-acid battery is made up of cells. Each cell is approximately 2 Volts. Therefore, a 12 Volt lead-acid battery has six individual cells. A fully charged cell has a voltage close to 2.12 Volts, while a discharged cell is close to 1.9 Volts.

If you choose to size a 12 Volt battery bank, then you may see voltage values ranging from 12.1V to 12.7V (depending on the stage of the charging process). This will show you that the battery bank is fully charged or finishing the charging process. If you notice voltages below 11.5V, your battery bank will be discharged entirely. You should stop discharging your battery at 12.1 Volts (50%).

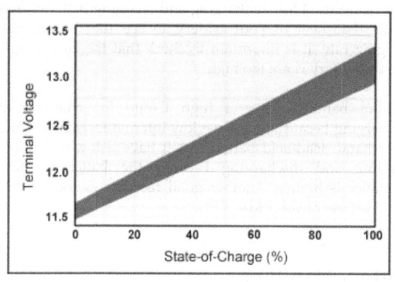

The state of charge vs. Terminal voltage
Source: Battery University

For a 24V lead-acid system, fully charged voltage is located at 25.5V, while a discharged battery is situated at 23V. Stop discharging at 24.2V (50%).

Finally, for the 48V lead-acid system, a fully charged battery bank is located at 50.9V, while a discharged battery is situated at 46V. Stop discharging at 48.4V (50%).

These voltages are temperature-dependent and model/technology-dependent; therefore, they can only provide a general reference.

The best way to check this is to verify the manufacturer's datasheet, which will likely include a voltage (V) vs. state of charge (%) graph.

Remember that no lead-acid battery should be discharged beyond 2V per cell or 50%. Otherwise, it could cause permanent damage to the battery.

As a reference, the following graph shows the percentage of charge with different lead-acid models.

State of Charge	One cell	6 Volts	12 Volts	24 Volts	48 Volts
100%	2.12	6.36	12.72	25.44	50.88
90%	2.08	6.24	12.48	24.96	49.92
80%	2.07	6.21	12.42	24.84	49.68
70%	2.05	6.15	12.3	24.6	49.2
60%	2.03	6.09	12.18	24.36	48.72
50%	2.01	6.03	12.06	24.12	48.24
40%	1.98	5.94	11.88	23.76	47.52
30%	1.96	5.88	11.76	23.52	47.04
20%	1.93	5.79	11.58	23.16	46.32
10%	1.89	5.67	11.34	22.68	45.36
0%	1.75	5.25	10.5	21	42

The State of Charge vs. Open Circuit Voltage

You can purchase battery monitors to tell you the battery's current voltage level. These monitors will only work in an open circuit, meaning no load or power source is attached.

The attached load will temporarily decrease the voltage at the battery's terminals, while the power source (PV panels) will increase the voltage at the terminals.

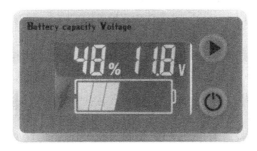

DROK battery monitor

Lithium batteries are different from lead-acid batteries. Each cell has a voltage of 3.2 Volts, and four individual cells make up one 12.8V battery.

Percentage (SOC)	1 Cell	12V	24V	48V
100% Charging	3.65	14.6	29.2	58.4
100% Rest	3.40	13.6	27.2	54.4
90%	3.35	13.4	26.8	53.6
80%	3.32	13.3	26.6	53.1
70%	3.30	13.2	26.4	52.8
60%	3.27	13.1	26.1	52.3
50%	3.26	13.0	26.1	52.2
40%	3.25	13.0	26.0	52.0
30%	3.22	12.9	25.8	51.5
20%	3.20	12.8	25.6	51.2
10%	3.00	12.0	24.0	48.0
0%	2.50	10.0	20.0	40.0

LifePO4 Voltage chart

Using voltage alone is not ideal for accurately measuring your battery's state of charge. However, voltage monitors are good if you are on a budget and have lead-acid batteries.

The discharge voltage graph of lithium batteries is very flat. The voltage doesn't change much between 10% and 90% state of charge. Therefore, it's hard to get a state of charge with a voltage meter if you use lithium batteries. If you use lithium batteries, you should always use a shunt. More on shunts at the end of the battery chapter.

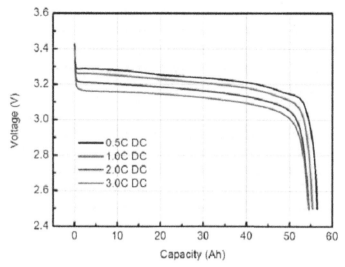

Voltage chart of a LiFePO4 battery at different C-rates

Sulfation

When you discharge a lead-acid battery below the recommended depth of discharge (50%), the sulfuric acid and electrolyte deplete. This effect creates large crystals of lead sulfate, which makes the charging and discharging process much harder and decreases efficiency.

Sulfation generally occurs after discharge at low currents due to acid stratification and crystallization. It also occurs when a lead-acid battery is stored for a long time in a discharged condition or never fully charged. That's why it's essential to have a trickle charger to keep the battery fully charged at all times.

Another cause is low electrolyte levels, which can be caused by excessive water loss from overcharging the battery or water evaporation and venting inside the batteries.

Sulfation is treated by charging the battery at a low current and a higher voltage (higher than nominal). This is generally between 2.4V and 2.5V and 0.5A to 8A per cell (depending on the battery size). In most cases, this will gradually reverse the sulfation process.

A sulfated battery can easily be recognized by looking at the plates inside it. A sulfated plate is lighter in color, and its surface becomes uneven and gritty. If you have a sealed lead-acid battery, it's best not to open the caps. Check the datasheet of your battery to learn more about maintenance.

Over-Discharging

If you're using lead-acid batteries, over-discharging can significantly shorten their lifespan, as these batteries lack a built-in battery management system (BMS). To prevent over-discharge, adding a low voltage disconnect (LVD) is essential if you're connecting loads directly to your lead-acid battery.

The LVD will disconnect the loads from the battery once it reaches a preset low voltage, preventing excessive discharge. Another option is using a shunt to monitor the battery's state of charge, though this won't automatically disconnect the load — it simply tracks battery usage.

The low voltage disconnect acts as a safety switch at a specific, pre-programmed voltage to prevent over-discharging.

Low-voltage protection is less of a concern for lithium batteries because they come with an integrated BMS that handles both low and high-voltage cutoffs automatically. The BMS will disconnect the battery when it reaches a low threshold, providing built-in over-discharge protection.

65 Amp low voltage disconnect from Victron

Overcharging

Overcharging occurs when a high charge voltage is applied to a battery even after it reaches 100% capacity. When this happens, excessive current flows into the battery, causing the water in the electrolyte to warm up, leading to premature aging. Overcharging can also cause corrosion of the positive plates, increased water evaporation, and, in severe cases, overheating. This overheating can lead to thermal runaway, a destructive process that can damage or destroy the battery within hours.

Preventing overcharging is essential for lead-acid batteries because they lack an internal battery management system (BMS). A charge controller is crucial in solar systems, as it regulates the current flow to the batteries, stopping the charge once the battery is full. This ensures safe charging and extends battery life. Selecting the correct charging profile for your specific lead-acid battery type within the charge controller is equally important to prevent overcharging.

Additionally, if you're using shore power or an external AC charger, be sure it has an adjustable charging profile for lead-acid batteries.

For lithium batteries, overcharge protection is managed by the BMS, which automatically regulates charging to prevent damage from excessive voltage. When you buy a lithium battery (LiFePO4) the BMS is already inside the battery case, so you don't have to add it yourself.

State of Charge

The state of charge (SOC) represents the available energy inside the battery at a specific moment. It's expressed in percentage.

When fully charged, the state of charge is 100%. The SOC depends on the consumption of the load and temperature values. The SOC can be figured out using the following methods:

- The easiest way to find out the SOC is to measure the battery voltage at the terminals. As mentioned in the book, it's best to measure voltage without a load.
- You can also check the voltage on the display of your charge controller.
- Use another instrument called a hydrometer. This device measures the specific gravity of the fluid inside a lead-acid battery.
- Read the battery capacity shunt monitor.

Depth of Discharge

The depth of discharge represents the amount of energy (in percent) of its total capacity extracted from the battery. DOD is the opposite of SOC, or in other words:

$$DOD = 100 - SOC$$

If your battery has a state of charge (SOC) of 80%, then the depth of discharge (DOD) is 20%. In other words, you have used 20% of the total capacity.

Maximum Cycles

The maximum number of cycles is related to the maximum number of charges and discharges applied to the battery. This value is highly dependent on the battery's DOD. For example, an AGM lead-acid model with 290Ah capacity from a manufacturer called Rolls gives a perspective of the variation of this parameter in the function of the DOD, as can be seen in the following figure.

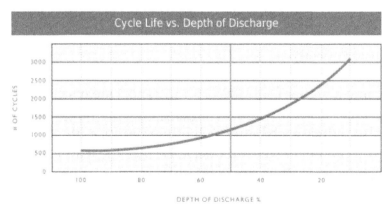

Cycle life vs. Depth of discharge
Source: Rolls

The nominal reference value in the graph is 50%. This reference value applies to all lead-acid batteries used for solar power applications. So, in this example, you can get 1,200 cycles at a DOD of 50% and 1,500 at a DOD of 40%.

When sizing your PV system, the fundamental consideration is to size the battery bank with a 50% DOD (for lead-acid).

If the system is designed to work daily, then the loads may draw a maximum of 50% of battery capacity daily. The solar panels need to recharge 50% of this capacity to reach 100% SOC the next day.

As shown in the previous figure, using a 50% DOD equals to 1,200 cycles of the battery, which means it can be charged and discharged 1,200 times.

The lower the DOD, the higher the cycles.
The higher the SOC, the higher the cycles.

50% DOD is used as a design parameter because it is the best trade-off between the number of cycles and the number of batteries needed.

Choosing a lower DOD (30%) means that your batteries will last longer. However, you will also need more batteries to cover the required capacity.

Choosing a higher DOD (80%, for instance) means you can take advantage of each battery charging and discharging cycle. It will also reduce each battery's lifetime significantly. Remember that you can never size a 100% DOD since the lead-acid battery will not recover all its capacity.

Lithium batteries today are built for exceptional longevity, with many cells capable of achieving up to 4,000 cycles. That's equivalent to about 11 years of daily cycling from 0% to 100%, after which the battery retains around 80% of its original capacity. If you cycle from 10% to 90%, you can reach 8,000 cycles and still have 80% usable capacity left.

This means the battery cells will likely outlast the battery management system (BMS). I expect the BMS lifespan to be closer to 10 years, similar to that of a string inverter. So, while it's perfectly fine to cycle a lithium battery fully each day, the BMS will likely need replacing long before the cells show significant capacity loss.

National Electrical Code

The National Electrical Code (NEC) 2017 (NFPA 70) is the most recent American reference for electrical installations, including photovoltaic systems and recreational vehicles.

The NEC 2017 is extensive and covers multiple typologies and electrical systems. For our scope of research, this part will focus only on the most important sections related to battery systems associated with recreational vehicles that you should consider when sizing and performing the installation.

Article 480 Storage Batteries — Section 4 (B)
Intercell and Intertier Conductors and Connections

This section refers to the requirements for series and parallel connections of batteries. As stated in this section: *The ampacity of conductors and connectors must have such cross-sectional area (gauge) that the temperature rise under maximum load conditions, and the maximum ambient temperature shall not exceed the safe operating temperature of the conductor insulation or the material of the conductor support.*

There are two ways of covering this requirement:

The first refers to the thermal equation of a conductor to calculate the temperature of the conductor for a given ampere value. This approach involves a quite extensive and complicated engineering equation as to be explained in this practical handbook.

The second option is to refer to the wiring chapter and look at the maximum amps for each cable diameter.

You must accurately estimate the maximum current load you will have in your system and make sure that this value does not exceed the permitted ampere capacity of the selected conductor. This is what was described as ampacity in the wiring section. The conductor heats up as higher electric currents pass through its cross-sectional area.

If you size your wire to withstand the ampere load demand of your system, then it will never surpass safe operating temperature values. Moreover, as stated in section 480.4 (B), you must also consider the maximum ambient temperature of the location. The procedure to calculate the influence of this parameter in cable sizing was explained in the section related to wiring.

Article 480 — Section 4 (C)
Battery Terminals

This section establishes that the connections to the battery terminals using cables cannot put a mechanical strain on them. This can be easily followed by properly sizing the cable and adjusting the length to give enough flexibility to the wires.

As stated in the section: *Electrical connections to the battery and the cable(s) between cells on separate levels or racks shall not put a mechanical strain on the battery terminals. Terminal plates shall be used where practicable. Informational Note: Conductors are commonly pre-formed to eliminate stress on battery terminations. Fine stranded cables may also eliminate the stress on battery terminations.*

Article 480.10
Battery Locations

This article focuses on establishing that the site for the battery location must have proper ventilation for the sufficient diffusion of gases from the battery to avoid the accumulation of gases that could become an explosive mixture.

Checking the Fire Code NFPA 1-2015, chapter 52 will give a further assessment of this aspect.

As stated in section 480.10 (C): *For battery racks, there shall be a minimum clearance of 25-mm (1 in.) between a cell container and any wall or structure on the side not requiring access for maintenance. Battery stands shall be permitted to contact adjacent walls or structures, provided that the battery shelf has a free air space for not less than 90% of its length.*

Another important consideration explained in this section is that the location for the battery bank must have a proper illumination source.

Article 551
Recreational Vehicles and Recreational Vehicle Parks

The National Electrical Code also considers the installation of recreational vehicles and park trailers as a separate electrical installation that has its own standards and requirements.

Article 551 exposes all the general requirements that must be carried out in electrical installations within recreational vehicles. Since there are too many requirements, we can only focus on the most relevant ones for solar power battery applications. However, if you intend to do your RV electrical installation as a DIY project, then you should verify article 551 of NEC 2017.

From this article, we can extract some relevant sections for this book.

Section *551.30 General Requirements* establishes that storage batteries must be secured in place to avoid any displacement from vibration and road shock.

Section *551.31 Multiple Supply Source* establishes that a multiple supply source (like a battery power station powered by solar energy) must have installed an overcurrent protective device (fuse) for the feeder of the alternate power source.

Article 552 Park Trailers

This article focuses on another type of vehicle but that can also be applied for recreational vehicles, cabins, or boats in solar power installations.

Section 552.10 (C) establishes the minimum separation requirements between the battery and other low-voltage circuits that must be physically separated by at least a 13-mm gap. The best way to ensure such a minimum gap will always be present. The best methods are to use clamping, routing, or any other equivalent means to guarantee a permanent separation.

This section also refers to the ground connections that must be done in an RV:

Ground connections to the chassis must be done in an accessible location and must be mechanically secure. Ground connections shall be using copper conductors and copper or copper-alloy terminals of the solderless type identified for the size of the wire used.

The surface on which ground terminals make contact must be cleaned and free from oxide or paint, and must be electrically connected through the use of cadmium, tin, or zinc-plated internal/external-toothed locking terminals.

Ground terminal attaching screws, rivets or bolts, nuts, and lock washers shall be cadmium, tin, or zinc-plated except rivets shall be permitted to be un-anodized aluminum where attaching to aluminum structures. The chassis-grounding terminal of the battery shall be connected to the unit chassis with a minimum 8AWG copper conductor.

Grounding in any PV system is one of the most important elements. Grounding provides a secure path for failure or short-circuits currents to flow to ground to avoid any possible damage to the components of the system. Following these considerations shall guarantee you will have a good grounding system.

Grounding doesn't necessarily mean connecting the vehicle to the ground. In an RV, the ground will be the chassis of your vehicle. Run a grounding wire from your inverter, charge controller, and battery negative to the chassis. These wires need to be separate. They cannot be placed in series with each other.

For example, one wire from the inverter to the ground of the chassis. The recommended wire size will be mentioned in the manual of the appliance. Then, another wire from the charge controller (if there is a connection point) to the same grounding point on the chassis.
Lastly, an 8-gauge copper wire from the negative battery terminal to the same grounding point on the chassis. More on grounding later in the book.

Finally, section 552.10 (D) also refers to the space and ventilation area of the battery bank. As stated in this section, the compartment where the batteries will be installed must be ventilated with a minimum opened area of 1100-mm^2 at both the top and bottom of the compartment.

In case the compartment doors are equipped for ventilation, the openings shall be within 50-mm (2 inches) of the top and bottom of the compartment. It is very important to consider that batteries cannot be placed in a compartment that shares space with a spark or flame-producing equipment.

Maintenance of Batteries

The maintenance schedule depends on the battery type you selected for your PV application. As we discussed earlier, lead-acid VLA batteries need more maintenance than VRLA batteries, and lithium batteries don't require any maintenance at all.

However, many steps in the maintenance procedure apply to lead-acid batteries.

- Check the battery's state of charge (SOC), as it will give you a reference for the current state of the battery before starting maintenance. This can be done by checking the charge controller's LCD screen. Another way is to use a voltmeter to check the voltage on battery terminals and verify the state of charge. Make sure the battery is fully charged before starting the procedure. Voltage has to be measured in an open circuit.

- Wear personal protective equipment such as protective eyewear and gloves when maintaining batteries. It's advisable to wear long sleeves and pants. I would even recommend a full face shield to ensure your whole face is protected. Safety glasses are still required underneath the face shield.

- The next thing to remember while maintaining your system is to disconnect the PV panels using the solar disconnect switch. Then, use the battery bank disconnect switch to disconnect the entire battery bank from the charge controller.

- Inspect the terminals, screws, clamps, and cables of the battery to verify if there is any damage or loose connections. These connections should be clean, tight, and free of any corrosion.

- Remove the battery from the compartment and place it over a non-conductive surface.

- The next step is to proceed to clean the battery. This can be done by using a mixed solution of distilled water and sodium bicarbonate (a proportion of 100 grams per liter). Using this mix, the battery case and terminals must be cleaned using a wet sponge (not dripping).

- Make sure the entire battery is thoroughly cleaned and free of dirt or grime.

- The next step depends on whether or not you have a vented lead-acid battery (VLA). VLA batteries need to be filled with distilled water periodically to recover their composition. When performing maintenance, open the valve and check the water level. Water should cover the battery up to the top of the cell. Use distilled water for this purpose. Always wear safety glasses and a face shield when performing this. You can also check the sulfation in this step.

- Make sure the battery is charged before starting the procedure. If not charged, fill the battery with enough water to cover the plates, charge it, and add the remaining water. VRLA batteries do not need this step.

- Put the battery back in place and check the connections properly. Loose connections can cause sparks or partial external discharges in the battery's terminals, which could be dangerous. The recommended torque for threaded battery terminals is 100-120 in-lb or 15 Nm, depending on the size of the terminals.

Complete battery maintenance should be performed every six months, at least for VRLA batteries. In the case of VLA batteries, maintenance may need to be carried out every month, especially checking the water levels.

For both types of batteries, checking the voltage, the ambient temperature of the compartment, the integrity of the battery system, and the ventilation check should be performed monthly.

If you have winterized your lead-acid batteries, remember to charge them once every month, even when not using them. This is because lead-acid batteries self-discharge around 15% each month, while lithium has a very low discharge rate of only 2% every month.

Buying Used Batteries

While buying used batteries might seem like a cost-effective option, it often comes with significant risks. Batteries are sensitive to various factors, including handling, maintenance, storage temperatures, and discharge rates, all of which can dramatically impact their life cycle.

One of the most significant issues with used batteries is the difficulty in determining how many cycles are left. The battery's age, usage patterns, and storage conditions aren't always verifiable, making it nearly impossible to assess its remaining lifespan accurately. While you can perform a capacity test by discharging the battery at a specified C-rate, this test is only reliable if the battery is fully charged and can be monitored under controlled conditions — a challenge if you're picking up the battery from a remote location.

Additionally, even if a used battery appears in good physical condition and charges or discharges as expected, there is no guarantee that it will continue to perform reliably over the long term. Batteries that have been improperly stored or allowed to self-discharge for extended periods may suffer from diminished capacity and efficiency.

Lead-acid batteries are particularly vulnerable to deterioration when used, as they do not have an internal battery management system (BMS) to prevent overcharging or over-discharging. Long-term exposure to deep discharges, incorrect charging, or high temperatures can permanently degrade their capacity, and there's no simple way to determine if a lead-acid battery has been well-maintained.

With lithium batteries, buying used also carries risks, though slightly less so if the battery has a functioning BMS. The BMS helps regulate charge and discharge limits, protecting lithium cells from common forms of damage.

However, while the BMS helps, it doesn't protect against all long-term degradation factors. For instance, lithium batteries that have been heavily cycled or exposed to extreme temperatures will have reduced capacity, even if they still charge and discharge normally.

A used lithium battery may also have a BMS that's nearing the end of its lifespan, as these systems often last around 10 years — considerably shorter than the potential cycle life of lithium cells, which can reach up to 8,000 cycles or about 22 years.

Used batteries — whether lead-acid or lithium — may also suffer from extended periods of self-discharge, which can cause permanent capacity loss, especially if left for months in a discharged state.

Even if a seller claims that the battery is "like new" or "barely used," it has likely been stored improperly, affecting its performance.

Do a discharge capacity test if you want to test your lead-acid battery. You can test the capacity of a battery, but you would need several hours or days to do it. This is how to do it:

1. Charge the battery to full capacity. For a 12 Volt lead-acid battery, this is 12.7 Volts.
2. Get a balance charger (Imax B6) or another brand. Select the Pb (lead-acid) or LiPo program from the menu.
3. Check that the battery discharge voltage is correct. This should be 12.1 Volts for a 12 Volts lead-acid battery (50%).
4. Select discharge current according to the C-rate.
5. Depending on your battery size, It will take a certain amount of time. For example, if you have a 100Ah lead-acid battery bank, it will take 50 hours at a one-amp discharge rate to get the battery to 50%.
6. The display on the balance charger should read half of the battery's capacity (50Ah). If it is less than half, the battery no longer has its true capacity.

If you buy used batteries, I recommend wiring them in parallel. This configuration minimizes the impact of capacity differences between batteries, allowing them to work together without reducing the overall capacity of the battery bank.

Weight of Batteries

Weight is an important factor when choosing batteries, especially in compact spaces like RVs, boats, or small cabins, where batteries might need to be mounted on shelves.

Lead-acid batteries are significantly heavier than lithium, which could influence your decision: a 12V 100Ah AGM battery weighs 64 lbs (29 kg), while a 12V 100Ah LiFePO4 battery weighs only 25 lbs (11 kg). Reducing weight can be beneficial in RVs and mobile setups.

Always secure batteries with straps, and consider using shock-absorbing mats to minimize vibrations if you place them in a vehicle. High-density foam can help with this.

Shunt

The shunt is a device that provides a visual and practical indication of the battery's state of charge.

Voltage-based battery monitors can accurately indicate lead-acid battery capacity. However, this is only true if no load is applied or if the battery is not being charged.

Unlike battery voltage meters, a shunt counts the number of amp-hours going in or out of the battery. This, combined with the battery's voltage, gives you Watt-hours. This is more accurate than the voltage meters.

As you already know, the battery's voltage drops once you apply a load to it. If you remove the load, the voltage goes back up.

Let's say the battery is at 80%, which is 12.42 Volts. If you apply a load, the voltage will drop. Because of the voltage drop, the voltage monitor will say 11.88 Volts, which is 40%, but in reality, it is 80%.

This is the same when charging the batteries. The batteries will indicate a higher voltage level when they are in an open circuit, which can be confusing. Therefore, a real-time battery capacity indicator, called a shunt, is needed.

There are several types of shunts, ranging from very cheap to expensive. The first one we will discuss is not recommended for installation in a solar system.

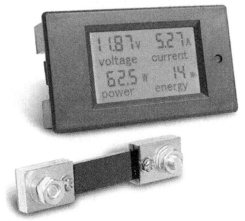

100A battery shunt

This shunt only measures in one direction. This means it only measures the energy going in **or** out. It is not a good representation of the current state of charge. This can be useful if you do not have solar panels installed. For example, you charge the battery in the RV park and monitor the drawn Watt-hours or Amp-hours over a few days until you reach another RV park. Then, you manually tell the shunt that the battery is fully charged. This is not useful if you charge the battery with solar panels during the day.

The other type of shunt measures both ways. This is much more interesting for solar applications.

This type measures both the current that flows into your batteries and the current drawn from them. It is recommended that you use this kind of shunt to monitor the true capacity of your battery bank. Victron makes one with Bluetooth capability so you can monitor your battery's capacity from your phone.

The one from Victron can be bought for $100 for the 500 Amps shunt with Bluetooth and another $110 for the display. Another one, sold by AiLi, is rated for 350 Amps and can be purchased for $45 with a display, but it's without Bluetooth.

Both meters require that you read the manual to set up the meter according to your battery pack.

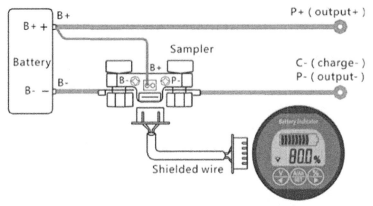

Wiring diagram for the Aili 350A shunt

It needs a positive voltage signal, which should be taken from the battery's positive terminal (or as close to the battery as possible). If you watch my videos, you can see that I have added a 1A glass fuse on the sampler wire. This is required because every wire in your system needs to be fused. The shielded wire goes to an easily accessible display.

The shunt is placed on the battery's negative terminal. From here, all the negative leads go to their destination. As can be seen in the following diagram, replacing the battery's negative terminal with the shunt is essentially the same.

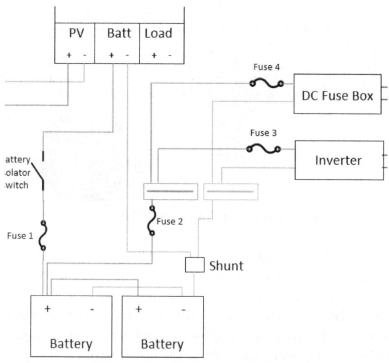

Schematic with integrated shunt

Solar Panels

Solar cells are the primary source of power in a PV system. The cells are made of silicon, which is the most abundant and economically attractive semi-conductive material (elements that can behave as isolating or conductive materials) to manufacture solar panels. Silicon is composed of electrons, neutrons, and protons, as any other element of the periodic table.

The process by which solar cells generate electricity is called the photoelectric effect, a physics phenomenon discovered by Albert Einstein. Briefly explained, the science behind it implies energy transformation from light into electricity.

Solar radiation has a broad spectrum. Based on this spectrum, solar radiation can be divided into two major components:

- Heat
- Light

The region of solar radiation wavelength that the solar panels can use for generating electricity is located within the visible light spectrum. Within this wavelength range, Sunlight particles have intrinsic kinetic energy that allows them to travel from the Sun to the Earth.

When these particles reach the surface of a solar cell made of silicon, they transfer this kinetic energy to the electrons of the silicon atom. This energy transfer makes the silicon behave as a conductive material and allows a small electric current flow.

Without entering into further physics concepts, the output of the solar cells can be combined through series/parallel connections to create a structure that we commonly call a solar panel or photovoltaic (PV) module. This allows us to increase the electric current, voltage, and conceivable power outputs used in common market applications.

Every cell you see on a solar panel can produce 0.5V. Adding 36 of these cells in series creates an 18V solar panel.

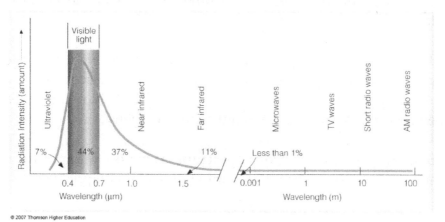

Solar radiation spectrum
Source: Atmos Washington

Types of Panels

The theory explained before applies to all solar panels. Some differences are worth noticing among PV modules. Thus, we can classify solar panels by technology.

Monocrystalline

Monocrystalline solar panels are the premium type of PV module available in the market. These modules have the highest light-to-electrical energy conversion efficiency in the market, with values that range between 19-24% (for recent top brands). That is why they are considered in many PV applications, especially those with little space available for placing solar cells.

Due to this premium performance, they also have a higher cost.

These PV modules are also requested for RVs, cabins, and boats since they optimize the space available to generate the highest electrical energy output. In addition, they are used in many home-type applications because of their elegant black or dark blue color. Monocrystalline solar cells usually have a rounded shape edge that is created due to the manufacturing process.

Monocrystalline solar panels
Source: Energy Global

The Czochralski method is used to manufacture monocrystalline solar panels. This process consists of melting multiple silicon rocks at 2,500°F (1,371°C) and dipping a silicon crystal seed into the melted solution.

As the crystal is slowly pulled upwards, a crystal structure is created around the seed, commonly called an ingot. The ingot is made with a cylindrical shape (the reason for the rounded edges) and sliced into multiple silicon wafers that are later transformed into cells.

The steps of the process can be seen in the following figure:

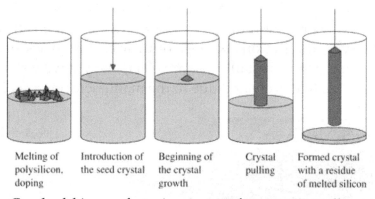

| Melting of polysilicon, doping | Introduction of the seed crystal | Beginning of the crystal growth | Crystal pulling | Formed crystal with a residue of melted silicon |

Czochralski manufacturing process for monocrystalline
Source: Top-alternative-energy-sources

Polycrystalline

The second option in terms of solar panel technology is the polycrystalline silicon module. These solar panels have lower efficiency values (between 16% and 19%) than monocrystalline modules. Still, their conversion efficiencies are good enough to be considered for the same applications as monocrystalline. Their most significant advantage compared to monocrystalline technologies is their price.

Polycrystalline solar panels are light blue in appearance. Depending on the brand and year of manufacture, granular shapes can sometimes be visible on the surface of the module. These modules look less elegant than monocrystalline technologies, which is why they are less requested when aesthetics is a must. Unlike monocrystalline panels, these modules have a squared edge.

Regarding the manufacturing process, these modules follow a procedure similar to that of monocrystalline.

However, a significant difference is that it is cooled down instead of pulling out the silicon crystal from the molted silicon solution. Then, the structure is sliced into multiple silicon wafers. The granular shape of some polycrystalline panels is due to the ingot being created from various silicon rocks.

Polycrystalline solar panels
Source: Solar Advice

A 100W polycrystalline panel will be bigger than a monocrystalline 100W panel. This is because polycrystalline has to account for lower efficiency. If space is a constraint, use monocrystalline. If space is not a constraint, use polycrystalline.

Thin-Film Technology

Thin-film solar panel technology is the market's third and last type of variation. These modules are made of incredibly thin films that are nearly 20 times thinner than the typical silicon-based panel, a property that makes them flexible and lightweight. If encased within plastic materials, the cells are flexible enough to adapt to the roof's shape surface.

Thin-film modules can be divided into four possible types:

- Amorphous-silicon (a-Si)
- Cadmium-Telluride (CdTe)
- Copper Indium Gallium Selenide (CIGS)
- Organic Photovoltaic (OPV)

Amorphous Silicon (a-Si)

The oldest thin-film technology is amorphous silicon. This type of module absorbs a wide range of light spectrums and is manufactured with non-toxic materials. Typically, small gadgets like solar calculators, watches, and outdoor solar chargers use a-Si cells because these devices need very low amounts of energy to work.

One of the most significant downsides of this technology is the efficiency values, which typically range between 10-13%, which is too low for residential or commercial applications. On the other hand, flexibility and cost are their biggest advantages.

The manufacturing process for these modules is different from that of their predecessors. Unlike polycrystalline or monocrystalline modules, amorphous solar panels are made from a thin plastic roll 30 micros thick, passing through a metal deposition machine that places a thin layer of silicon material onto the plastic. Then, another machine uses lasers to scribe the material intersections that define the individual solar cells.

Thin-film roll of solar panels
Source: Clean Energy Authority

Cadmium Telluride (CdTe)

Cadmium telluride is the most common type of thin-film technology used in commercial applications. The leading company is First Solar, which dominates the utility-scale sector. These modules' conversion efficiency values from First Solar can range from 15% (Series 4) up to 18% (Series 6), suitable for utility-scale applications.

CdTe panels have better efficiency values related to lower light wavelengths and can be manufactured at lower costs since cadmium is abundant as a zinc byproduct. However, the main disadvantage of CdTe modules is pollution since cadmium is a highly toxic material.

Although using these modules in residential or commercial applications is not dangerous for human health, the recycling process of these panels is another matter.

First solar PV modules installed in a 40-MW power plant
Source: Renewables Now

Copper Indium Gallium Selenide (CIGS)

CIGS modules are generally produced through co-evaporation or co-deposition techniques. Copper, indium, gallium, and selenide are placed on the substrate (plastic, steel, glass, or aluminum) at different temperature rates and sandwiched between conductive layers. When placed on a flexible backing, the layers are thin enough to bend at the user's will up to a certain point.

CIGS module manufacturers such as Sunflare, MiaSolé, and Solar Frontier typically focus on markets that silicon-based technologies cannot cover. For instance, MiaSolé focused on commercial rooftop applications. However, they shifted the market sector to transportation and trucks to provide an environmentally friendly fuel consumption reduction solution.

They also manufacture flexible solar cells placed on a steel substrate with efficiency values reaching 17%. Their modules can be installed through a peel-and-stick system, making them easier and cheaper to install on trucks and carports. These properties make Sunflare, Solar Frontier, MiaSolé, and other CIGS module manufacturers ideal for RVs, vans, and boats with curved surfaces.

The main disadvantages of CIGS modules compared to CdTe panels are the price and the reduced capability of heat dissipation, which we will discuss soon.

Sunflare flexible solar panels on a curved trailer
Source: Sunflare

Organic Photovoltaic (OPV)

The organic photovoltaic panel is made from conductive organic polymers that generate an electric current after depositing multiple layers of thin organic vapor between two electrodes.

These solar cells are ideal for new applications such as building-integrated photovoltaics (BIPV). Thanks to their ability, OPV panels can be colored in several ways or even made transparent. This is perfect for BIPV applications, where color variation is an excellent addition to integrating solar panels into building windows. Due to the abundance of organic polymer materials, manufacturing costs are low. Organic solar cells are also thin, flexible, and printable.

The main downside of this technology is efficiency since organic solar cells generally reach values close to 11%, well under the current market standard. Another issue is related to lifespan. Organic degradation does not occur in other technologies and reduces the years the cells can work efficiently.

Unique bus stop using organic solar cells

Other Cell Types

Gallium Arsenide (GaAs)

Another type of solar cell is the gallium arsenide solar cell, which has the highest efficiency values.

Alta Devices, a GaAs solar cell manufacturer, has developed a 29% efficient cell known as the dual-junction cell. GaAs cells also have other advantages: flexibility, lightweight, adjustability to multiple colors, thinner and malleable structure, good temperature resistance, and good performance under low light conditions.

Despite this high-efficiency value, GaAs cells have a significant disadvantage: high costs.

Since gallium is scarce and arsenic is toxic, these solar cells' raw materials and manufacturing process costs are much higher than traditional silicon-based technologies. This is why GaAs is used in small applications where efficiency is crucial, like space aircraft applications. You are more likely to find GaAs solar cells than solar panels.

Dye-Sensitized

The dye-sensitized solar cell is based on a semiconductor generated between a photo-sensitized anode and an electrolyte. These cells are easy to manufacture through printing cell techniques. They are semi-transparent. Overall, conversion efficiency rates are close to 11%. These cells also work well under low light conditions.

Their main disadvantage is cost, which is why they are not used for residential, commercial, or large-scale power plants. As a result, they are the most forgotten solar cell in the market.

Perovskite

Finally, the ultimate type of solar cell technology is perovskite. Although not yet manufactured on a large scale, if it is successfully deployed in the next few years, this technology is expected to revolutionize the solar panel manufacturing industry.

Efficiency values are expected to be at least 25% and may reach 30%. In addition, low manufacturing costs, flexibility, and printability make it an attractive option for future market development. However, it contains lead, which strains the public's acceptance of this solar panel.

Printable perovskite solar cells
Source: Instyle Solar

Now that we have discussed the different solar panels, we will discuss their characteristics, including series and parallel, tilting, shading, and more.

Conversion Efficiency

The conversion efficiency of a solar panel represents the maximum power output that the module can provide based on a specific module size area. Therefore, a solar panel with higher efficiency needs less space to give the same power output.

The maximum theoretical value that a silicon solar panel (based on a single-junction structure) can achieve is 33.7%, which is the Shockley-Queisser limit.

In other words, from all the power in sunlight (roughly 1000 W/m^2) hitting an ideal solar cell, only 33.7% can be converted into electricity — about 337 W/m^2.

I-V Curve

Since solar panels generate DC electricity, two parameters determine the power output of the PV module:

- Voltage
- Current

As you already know, voltage (V) multiplied by current (I) makes up a device's power (Watt).

As I will explain later, voltage and current parameters vary according to ambient conditions. The pattern change of these two parameters follows a specific curve. This curve aims to determine the equivalent power output for two voltage and current values provided.

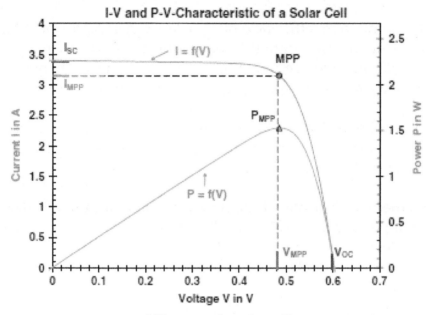

I-V curve of a solar cell
Source: H.Haberlin
"Photovoltaics — System Design and Practice"

If we look closer, there are two lines. The line on top: I=f(V) represents the I-V line that shows variations of current according to voltage values. The I-V line shows multiple parameters that are worth noticing.

Voc and Isc

We must first notice the open-circuit voltage (Voc) and the short-circuit current (Isc). These parameters are located on the external points of the line, and they represent the highest values that both voltage and current can have.

Specifications of a solar panel
Source: Sunpower

To understand how this curve works, we can position ourselves on the highest point of the curve, which is Isc. This point represents a short-circuit condition in which the solar panel is connected to a very low resistance (ideally zero) that allows electrical current to flow at maximum value. It would be equivalent to wire the positive and negative terminals of the panel together.

When the resistance is increased, the voltage starts rising. The current starts reducing step by step until resistance is too big to allow the current to flow, which leads to the open-circuit condition.

Under this condition, voltage is at its highest value (Voc), and the current is zero. This is equivalent to leaving the two terminals of the solar panel without connection to any load (here, the maximum resistance is the non-conductive air).

Impp, Vmpp, and Pmpp

The following two parameters on our list are the maximum power point current (Impp) and the maximum power point voltage (Vmpp). Sometimes Imp and Vmp are used, which are similar. These two points are linked directly to the fifth point of the curve, the maximum power point (Pmpp or simply MPP).

The MPP represents the maximum power output that the solar panel can provide for specific ambient conditions. Vmpp and Impp represent the corresponding voltage and current values (respectively) associated with the MPP point.

$$Pmpp = Impp \times Vmpp$$

Calculating the Pmpp or Wp (Watt peak):

$$5.80A \times 17.5\,V = 101.5 Watts$$

MODEL: SPR-E-Flex-100		
Rated Power (Pmax)[1] (+/–3%)	100	W
Voltage (Vmp)	17.5	V
Current (Imp)	5.80	A
Open-Circuit Voltage (Voc)	21.0	V
Short-Circuit Current (Isc)	6.20	A
Maximum Series Fuse	15	A

Standard Test Conditions: 1000 W/m², AM 1.5, 25° C
Suitable for ungrounded, positive, or negative grounded DC systems
Field Wiring: Cu wiring only, min. 12 AWG/4 mm², insulated for 90° C min.

$C \in$

RoHS

Pmpp, Vmpp, and Impp of a solar panel

P-V Curve

The line on the bottom, P=f(V), is known as the P-V curve. It represents variations of power output with respect to voltage. Here the Pmpp (MPP) is the only point of interest. The linear relation between current and voltage can be seen until MPP is reached.

STC and NOCT

There are hundreds of solar panel manufacturers available on the market. Therefore, the solar industry needs a way to categorize and compare modules. This is done through a laboratory test under which all solar panels must be submitted to test their performance under the same conditions. These are known as the Standard Test Conditions (STC).

Standard Test Conditions: 1000 W/m², AM 1.5, 25° C

STC on a solar panel

The STC reference parameters used in lab tests are:

- Irradiance: $1kW/m^2$
- Temperature: 25°C (77°F)
- Air Mass: 1.5AM

This temperature is referenced to the module's operating temperature (not ambient temperature). All parameters explained before in the I-V curve will be referenced to STC in the solar panels' datasheet.

Another typical reference value is the NOCT, the acronym for Nominal Operating Cell Temperature. This standard uses parameters closer to the typical solar panel operation since STC conditions are often unreal. The temperature value that is stated in NOCT represents the temperature of the cell under the open-circuit condition and the following circumstances:

- Irradiance: $800W/m^2$
- Wind Speed: 1 m/s
- Ambient Temperature: 20°C (68°F)
- The temperature on the surface of the panel: 45°C (113°F)
- Mounting system: Open rack

As we can see, the ambient temperature is different from the cell's operating temperature. Depending on the manufacturer, the NOCT temperature value will generally be between 45-48°C (113-118°F).

Effect of Insolation and Temperature

As we mentioned before, the I-V curve depends on ambient conditions, mainly on irradiance and temperature.

A higher irradiance (sunshine) means more solar radiation. Higher solar radiation also means more photons that reach the surface of the module and, therefore, more moving electrons. Since the displacement of electrons is linked to the flow of electric current, more electrons moving means higher current.

In other words, more solar irradiance means more current, and less irradiance means less current. The relationship between these two variables is proportional and linear. Irradiance (sunshine) does not affect voltage.

On the other hand, the temperature is different. Temperature affects all variables. However, the most important effect is on voltage. Unlike irradiance, the relationship between temperature and voltage is inversely proportional and logarithmic.

This means that when the cell's temperature increases, the voltage reduces, while if the temperature decreases, the voltage rises. The following figure shows a graph illustrating the effects of irradiance and temperature on current and voltage, respectively.

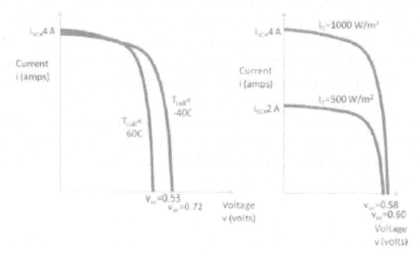

Effect of temperature on voltage (left)
Irradiance on current (right)
Source: A. Walker.
"Technologies and Project Delivery for Buildings"

On the left, the solar cell's voltage decreases with increasing temperatures while the current stays unchanged.

On the right, the current decreases once less irradiance (sunshine) reaches the panel while the voltage stays unchanged.

What we can learn from this is that we need to keep our panels as cool as possible while maximizing the exposure to the sun.

Ambient Temperature and Cell Temperature

The cell temperature increases according to two factors:

- The amount of current flowing through the cell.
- The ambient temperature.

The first one depends on the load that the solar panel is connected to and the irradiance levels. When current flows through any conductor, an ohmic loss effect is created, translating into heat. The same happens inside the solar cell. The second factor is dependent on the location where the panel will be installed.

As you can imagine, hot ambient temperatures will add a thermal effect to the module, increasing the cell's temperature. This is an undesirable condition as excessive temperatures decrease voltage and thus reduce the modules' power output.

On the contrary, low ambient temperatures favor the thermal cooling of the cell due to ohmic effects. Therefore, cool temperature locations are always desirable for solar panels.

Ironically, many locations with excellent solar irradiance also have high temperatures that translate into thermal losses (one of the most important photovoltaic losses). Therefore, in some cases, a location with a cooler ambient temperature and lower solar irradiance could be better for solar since thermal losses will be lower. You can increase the cooling effect by mounting your solar panels on a ground mount where circulating air can cool the panels underneath.

Temperature Effects on Efficiency

As stated before, temperature affects the solar panel's power output. As we mentioned in the conversion efficiency section, the solar panel's efficiency depends on the Pmpp. Therefore, temperature intrinsically affects the solar panel's efficiency as well. The relationship of this effect is linear, as can be seen in the following figure. This is an example of a solar panel with an efficiency of 14.8% under STC 25°C (77°F).

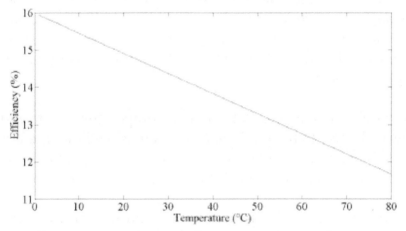

Efficiency variations according to temperature changes.
Source: "The Effect of Temperature on Cell Efficiency"

We can see that efficiency drastically decreases as the temperature increases. Therefore, your solar panels must receive as much ventilation as possible.

Series and Parallel Connections

Solar panels have specific power outputs in their datasheets. Despite new models that can reach values close to 400Wp, this power output is insufficient to cover the energy needs of household appliances or bigger systems. Therefore, solar panels need to be combined to increase the power output.

Series connections are the same as batteries and consist of connecting the negative terminal with the positive terminal of the next solar panel.

Meanwhile, solar panels connected in parallel consist of combining the positive terminals and the negative terminals in the combiner box or by using branch connectors.

A set of solar panels connected in series is known as a string. A mix between solar panels in series and parallel connections is known as an array.

When solar panels are wired in series, the voltage of each module is added while the current stays the same. For instance, if the solar panel's output is 10V and 1A, and you connect three modules in series, then the system's output will be 30V and 1A.

On the other hand, when solar panels are wired in parallel, the current increases while the voltage stays the same. So, based on the same example, if you connect three solar panels in parallel, the system's output would be 10V and 3A.

When multiplying the voltage by the current, both systems will provide 30W. So, the total power of these three modules does not differ if they are wired in series or parallel.

When we connect panels in parallel, the current increases. Higher current values translate into bigger gauges for PV wires. Therefore, to reduce costs, series connections are preferred when connecting solar panels.

As a general rule of thumb, solar panels must be wired in series until the accumulated voltage is under the maximum input voltage of your charge controller. The following example is a 3S setup (3 panels in series).

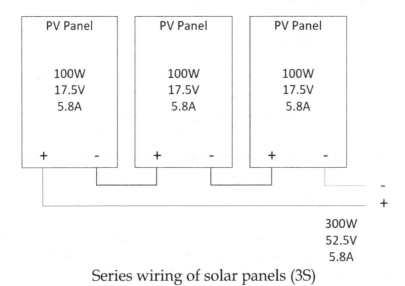

Series wiring of solar panels (3S)

The following example is a parallel connection. This setup is called 3P (3 panels in parallel).

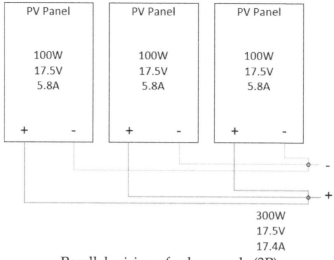

PV Panel	PV Panel	PV Panel
100W	100W	100W
17.5V	17.5V	17.5V
5.8A	5.8A	5.8A

300W
17.5V
17.4A

Parallel wiring of solar panels (3P)

Whether you place your panels in series or parallel will depend on the charge controller you use. A PWM charge controller will only take voltages as close as possible to the battery voltage, whether it be 12 or 24 Volts, while an MPPT can take voltages up to 100 Volts or more.

As you will learn later in the book, PWM charge controllers are cheaper than MPPT. If you wire your panels in series, the voltage will increase while the current stays unchanged. This will influence the diameter of your wire. The money you save on wiring in series instead of parallel can be spent on a more efficient MPPT inverter.

Another point to consider is the angle of the sun in the morning and evening. Because of the low angle, your panels won't generate as much voltage. If you have three panels in parallel and each one generates 5 Volts, you will send 5 Volts to your charge controller, which won't be enough to charge batteries (under the minimum required input voltage).

If you wire the same panels in series, you have 15 Volts (5V+5V+5V=15Volts), which can start to charge a 12V battery early in the morning or late in the evening.

It is worth noting that Y-branch connectors have a current limit. Most of them are limited to 20 or 30A, so ensure you do not exceed this current.

Y-branch connectors for parallel solar panels

Wiring Different Solar Panels

Another important rule that must be considered is that solar panels with different specs must never be wired together. Once you have selected a solar panel, you must purchase all the required modules for the PV system with that specific model.

You cannot wire solar panels with different specs because the PV system will not work optimally. The current output must be the same throughout the entire system in the series connection.

If four solar panels are wired in series, and one solar panel's output is 2A while the others are 3A, the connection will only provide 2A.

The solar panel with lower output would not be capable of providing 3A. Therefore, the system must adjust and deliver 2A. This translates into underusing the capacity of the other solar panel(s). A similar problem occurs with parallel connections but with voltage. If you have no other choice and you must connect mismatched panels, use this rule:

- Panels with the same voltage: wire in parallel.
- Panels with the same current: wire in series.

Solar Panel Array

Another factor should be considered when making series and parallel connections.

Let's say you have 8 solar panels. Due to charge controller voltage input restrictions, the maximum number of modules that can be connected in series is 5.

Now, you may think you could make a string of 5 solar panels and another string of 3 modules, connect the outputs in parallel, and wire it to the charge controller. However, this would be incorrect for two reasons.

The first reason is that the weaker string will have a lower voltage. This is because current always flows from the highest voltage point to the lowest voltage point.

This principle will generate an effect in which the other strings try to make the current flow toward the weaker string.

This is highly undesirable since it can lead to malfunctioning and be devastating under short-circuit conditions.

The second reason is related to energy losses. As will be explained later, the charge controller must accurately find the MPP of the solar panels every time to operate optimally. If strings of different voltages are connected in parallel, then the I-V curve will lose its regular shape, making tracking of the MPP very hard for the charge controller. This would result in mismatch losses due to voltage differences. This will be explained in further detail in the charge controller chapter.

Therefore, going back to our example, if you have 8 solar panels, you would have to size 4 modules in series (string) and put them parallel to become an array. This system is a 4S2P (4 panels in series with 2 parallel groups).

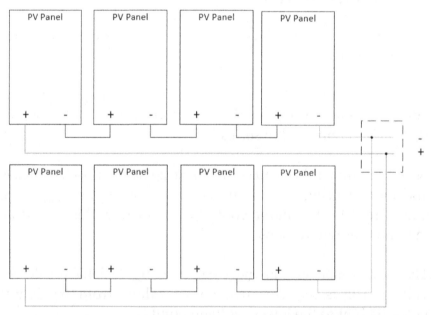

Two series strings connected in parallel to the combiner box (4S2P)

If the charge controller only allowed for 3 modules in series, you would have to either work with 6 panels or add an extra solar panel to have 9 modules in total, of which 3 are in series. This setup is called 3S3P (3 panels in series and 3 parallel groups).

As previously mentioned, ensure the current flowing in the array is suitable for the wire according to the ampacity and the voltage drop. Refer back to the wiring section to learn how to calculate both.

Azimuth

The azimuth angle refers to the direction of solar panels regarding the sun's orientation. Solar panels must face south to harness the maximum power output for locations in the Earth's northern hemisphere. For locations in the southern hemisphere, solar panels should face North. Locations near the equator should face the panels almost horizontally.

For U.S. cases, a solar panel should take south as reference 0°. If you install solar panels on RVs or boats, it makes little sense to determine the azimuth angle since the solar panels will vary their azimuth as they go on the road or sea. Also, if solar panels are mounted to the RV roof in a flat position, you shouldn't worry about that.

If the solar panels you have installed can be lifted with a certain tilt for maximum solar power harnessing once you park the RV, then parking the RV so that the solar panels would face 0° south (or as close as possible) would be beneficial.

Finding the optimum azimuth angle will be beneficial if you have a portable solar panel with the RV. For this, you must use a compass and place the panels facing south directly if you are not planning on moving the panels throughout the day.

Alternative azimuth directions are east or west. Panels should never point north if you are in the northern hemisphere.

Tilt Angle

The tilt angle is another important factor in solar power harnessing. Finding the optimal tilt angle is always related to the location's latitude.

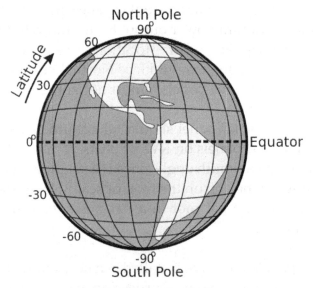

Displaying latitude
Source: Geography Realm

For locations near the Equator, choosing the latitude as the reference is usually the best approach.

Besides the location, another important factor that must be considered before setting the optimum tilt angle is the type of system that will be implemented. Since the altitude and direction of the sun vary according to the season, it is important to know when the system will be used most.

These are the formulas we use when calculating the optimal tilt angle:

- Winter: (latitude × 0.9) + 29 degrees
- Summer: (latitude × 0.9) – 23.5 degrees
- Spring and fall: latitude – 2.5 degrees

Here is how you can find out the latitude of your location:

Go to Google Maps and click on the location you would like to know the latitude for. The first number is the latitude. The one after it is the longitude. In this example, the latitude is 33° (rounded).

1654-1608 Secretariat Ln
Irving, TX 75060, USA
32.796378, -96.988878

Figuring out the latitude of your location

Let's apply this latitude to our formulas:

- Winter: (33° × 0.9) + 29 degrees=58.7°
- Summer: (33° × 0.9) – 23.5 degrees=6.2°
- Spring and fall: 33° – 2.5 degrees=30.5°

The spring and fall angle (year-round angle) will yield the most energy. I have made a YouTube video about this. Search for 'tilt angle', or go to https://cleversolarpower.com/book/tiltvideo to see several case studies.

Don't want to use the formulas? Use my online tilt calculator on my website: https://cleversolarpower.com/book/tilt

Shading

Shading losses are among the most underestimated factors in any PV system and must always be considered. There are two main types of shading: near-shading and far-shading.

Far shading is associated with losses in diffused irradiance caused by mountains or high buildings. Unfortunately, little can be done about this.

On the other hand, near-shading is associated with nearby objects that can project shade over the solar panels. For example, trees, walls, antennas, or an RV's air vent can create shade on the solar panel.

When a solar panel is shaded, the module's current output is affected since the obstruction reduces the number of photons that the module can absorb. In addition, the power output of the entire string of the shaded solar panel is affected because the electrical current that flows through a string (series connection) must be the same in every module.

If you remember the effects of solar irradiance (Watts/m²), you will know that it affects the current, and the temperature affects the voltage. So, if you shade one panel in a string, only the panel with the lowest current would decide the power output.

This can be seen in the following image.

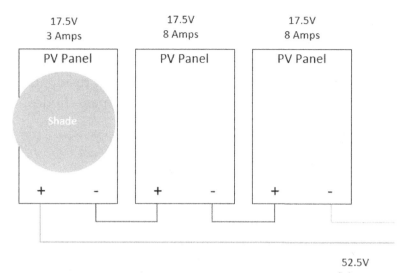

Effect of shading on series connections

The electrical current can only be as high as the current generated by the weakest module (shaded module). Therefore, the output of this string is only 3 Amps.

You should avoid any near-shading that could cause essential power losses to your PV system.

Solar panel manufacturers install bypass diodes in a box located in the back of the module, known as the junction box. If the solar panel is partially shaded, these bypass diodes allow electric current circulation from the other sides of the module.

This means the reduction in power output due to shading will not be total but partial. The following figure shows the structure division of a Panasonic solar panel with 4 bypass diodes. When a leaf partially shades one section, the bypass diode of that section activates to allow current circulation from the remaining solar cells. Most solar panels, in residential and commercial applications, generally have 3 bypass diodes.

Bypass diodes in solar panels
Source: Panasonic HIT module brochure

The trick lies in using this property of solar panels in favor of shading. By placing the module vertically or horizontally, the effect of shading can be very different.

Most modern panels have these bypass diodes. Check the panel's datasheet to ensure they have bypass diodes before buying the modules.

In the following diagram, you can see the effect of shading on parallel connections. We can see a higher power output if we compare this to the series connection. This is because, in parallel, the current is added together while the voltage stays the same.

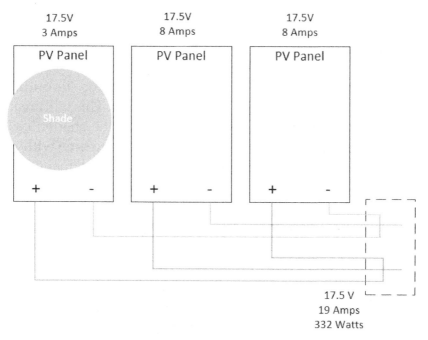

Effect of shading on parallel connections

We can see that three 140-watt panels in parallel will have a combined output of 332 watts instead of 420 watts. If we had the same setup in series, the output power would be 157 watts.

The downside of parallel connections is that you need a bigger wire diameter to handle the increased current.

To sum up, if you expect shade, it's best to wire in parallel. If you do not expect shading, wire the panels in series. Always choose a location without shading throughout the day. If your panels are on a ground mount, have at least one foot of clearance from the base of the panel to the ground.

Hotspots

You already learned that shading a panel will reduce its output. The energy that is lost will be dissipated as heat. This will create a hotspot on your panel. A hotspot is where the shaded cells dissipate the heat. If the hotspot is there for a long time without cleaning the panel, it can lead to permanent damage and reduced panel output.

Bird droppings, dust on the panel, or a plant's shade can cause hotspots, which can be detected with a thermal imaging camera.

Examples of hotspots
Source: review.solar

It is essential to avoid shading on a solar panel at all costs. This will reduce output drastically. Periodically clean your panels because the accumulated dust will reduce the power output. More about the effects of cleaning your panels later.

Blocking Diodes

Blocking diodes are sometimes used in battery-based applications that involve solar panels.

The current flow in any electrical system always goes from the highest to the lowest voltage point. Keeping that statement in mind, during the day, solar panels have higher voltages than batteries. Therefore, voltage naturally flows toward the battery to charge it.

During the night, there is no sunlight, and solar panels generate no power at all.

The battery will never be discharged entirely (or at least it shouldn't). As the only source of power in the PV system, the battery will provide electrical current to any other device. The electrical current could flow back to the solar panels during this time, making you lose energy.

When charge controllers did not exist, installers needed to add a blocking diode between the module and battery to avoid this reverse current effect.

Nowadays, manufacturers address this element by adding a Schottky diode, which combines the functions of blocking and bypass diodes. These diodes are already integrated into the charge controller.

Fusing Solar Panels

A short-circuit condition in solar panels can occur due to metal-to-metal contact of the PV wires in one of the strings due to a mechanical accident or lightning. Under this condition, the faulted string would receive the short-circuit current from every string of solar panels known as the reverse current (Ir).

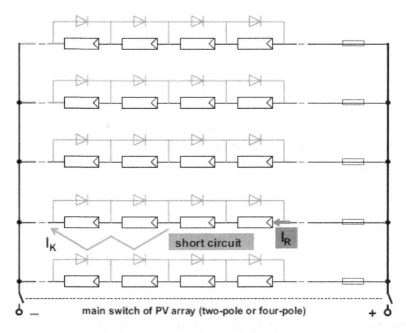

Short-circuit condition for a 5 strings PV system
Source: H.Haberlin
"Photovoltaics - System Design and Practice"

As shown in this image, in a PV system with 5 strings of modules (4S5P), the total short-circuit reverse current the faulted string could receive, will be four times the short-circuit current (Isc) of the module (close to 9A). In this case, the faulted string would receive a 36A short-circuit current. This would be destructive for both the modules and possibly the wires and could even induce fire.

Solar panel strings should be protected against high reverse currents, which can damage the modules and the PV wires in a short-circuit condition.

To protect the PV modules, DC-compatible fuses are installed on the positive side terminal of each series string. DC fuses act as overcurrent protection devices, isolating the faulted circuit from the rest by melting down a conductive material inside the fuse when a specific current passes through it.

These fuses are generally installed with an inline MC-4 connector or in a DC combiner box.

To select the string fuse, two factors must be considered:

- The open-circuit voltage of the module.
- The short-circuit current of the module.

It is crucial to size the string fuse for both factors. Only sizing the fuse as short-circuit protection would be unsuitable and could even cause malfunction and fire.

Since the DC signal never crosses through zero volts, safely isolating the circuit from the rest is much harder than with AC power. Therefore, you must ensure that the fuse you select has been designed for DC connections to interrupt the current flow safely.

Under specific weather conditions, solar irradiance values could be close to $1,000W/m^2$, which will cause additional stress to the fuses due to increased heat. Moreover, the fuse could be submitted to this stress with an additional rated maximum current flow, increasing the heat inside the fuse.

Assuming these conditions, security factors must be applied to size the fuse. A 25% security factor should be used due to excess irradiance and another 25% due to 3 continuous hours of operation under these conditions. Based on these considerations, the rated current for the DC fuse should be calculated as demonstrated in the following expression.

$$Ifuse = Isc \cdot 1.25 \cdot 1.25 = Isc \cdot 1.56$$

On the other hand, considering the open voltage of the modules, string fuses should be rated for 1.2 times the STC open-circuit voltage of the entire string. This voltage can be calculated by verifying the open-circuit voltage of the module model and multiplying it by the number of solar panels (n_{sp}) in every string. The result should be your minimum voltage to the DC fuse or breaker.

$$Voc_{stc-string} = Voc_{stc-modules} \cdot n_{sp}$$

$$Vfuse = 1.2 \times Voc_{stc-string}$$

Let's explain this with an example. Three panels in series with the following specifications:

MODEL: SPR-E-Flex-100

Rated Power (Pmax)[1] (+/–3%)	100	W
Voltage (Vmp)	17.5	V
Current (Imp)	5.80	A
Open-Circuit Voltage (Voc)	21.0	V
Short-Circuit Current (Isc)	6.20	A
Maximum Series Fuse	15	A

Minimum current for the fuse:

$$Ifuse = 6.2A \times 1.25 \times 1.25 = 6.2A \times 1.56 = 9.67 Amps$$

The solar panel specifications state that the maximum fuse in series is 15A. Do not use a higher fuse than the manufacturer recommends. You can either use a 10 Amp fuse or a 15 Amp fuse.

Minimum voltage for the fuse:

$$Voc_{stc-string} = 21.0 V_{stc-modules} \times 3_{sp} = 63\,Volts$$

$$Vfuse = 1.2 \times 63_{stc-string} = 75.6\,Volts$$

DC fuses are rated for a specific voltage. Choose a DC fuse that can handle at least 75.6 Volts. Most fuses rated for PV are 1000VDC.

Series Connection

For solar panel series connections, you don't need an inline fuse because the current remains the same across the series string, keeping it well within safe wiring limits.

Instead, a DC breaker or solar disconnect can isolate the solar panels from the system when needed. Note that this breaker serves as a switch for isolation, not as cable or PV module protection, as the current won't be high enough to risk overheating the wires. For parallel connections, however, fuses are essential to protect each string.

Using a DC breaker as an isolation switch

Parallel Connection

You can use inline fuses for parallel connections. However, most parallel systems use a combiner box. Combiner boxes combine multiple wires into one wire that goes to the charge controller. Wiring fuses this way can be cheaper than buying inline MC-4 fuses because you will need the combiner box anyway.

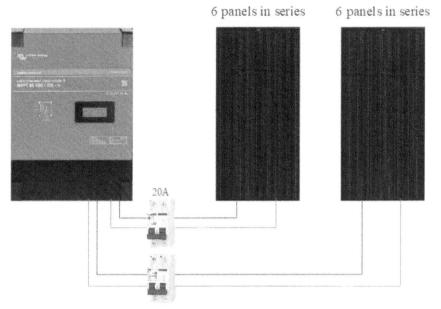

6 panels in series   6 panels in series

20A

Wiring DC breakers in a 6S2P parallel connection

In parallel connection, you need to fuse every series string. If you have 12 panels and configure them in a 6S2P setup, you have to use two fuses or two breakers (shown in the image). One protection for every series string. One fuse will protect 6 panels because the current in a series string is the same.

Example of an inline MC-4 connector fuse

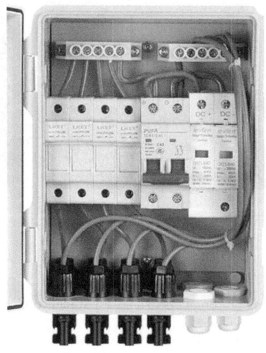

Combiner box from left to right:
4 fuses, main breaker, overvoltage protection from lightning

In my opinion, it's better to use a DC breaker instead of an MC-4 fuse. The DC breaker will act as a short circuit protection and a switch to isolate the solar panels simultaneously.

If you only use MC-4 fuses, you still need an additional PV disconnect switch.

Placement of Panels

When installing solar panels, you will always have two options if the system is intended for residential applications.

The first option is to install solar panels on your roof, and the second is to use ground-mounting.

Choosing a ground-mounting installation has several advantages. Orientation and tilt angle variability are much more flexible with ground-mounting than with roof mounting. Also, maintenance procedures are easier since the modules are placed in an accessible location. It is very important to clean the modules of dirt or snow.

These systems have a better cooling system since air circulation on the backside is better than roof mounting. You won't need to modify your house to install the modules, and from an expansion perspective, ground mounts are much easier to expand.

On the negative side, ground mounts are generally more affected by shade than roof mount types since more objects can project shade over the modules. Besides, the installation procedure is often more complicated and probably more expensive than that of a roof mount. Finally, another important point is that you will reduce space in your backyard that could be used for recreational purposes.

If you choose to go with roof-mounting for your home, you will also have some benefits. Among them, installation costs will be lower (unless your roof requires structural updates).
The solar panels can also protect your roof surface from hailstorms since they are made to resist impacts (to a degree). The low accessibility to your solar panels may be bad on one side from a maintenance perspective. However, from a security perspective, your solar panels could rarely be stolen or even vandalized since they are on the roof.

Meanwhile, installing a roof-type solar system will require you to take care of several details, such as roof penetration during the installation, maintenance, limitations on system size due to space availability, and roof structure upgrades if needed to hold the weight of the modules.

Based on these pros and cons, you will need to balance and decide where to place your solar panels, either on the roof or on the ground. If you have the space, I recommend using a ground-mount system.

Mounting Panels

The module mounting procedure depends on the mounting system you would like to use.

Roof-Mount

For roof-mounting systems, the following procedure must be followed:

1. You need to have your equipment and materials — drilling, pencils, chalk blocks, rails, clamps, bolts, and screws.
2. Calculate the distance between rails on the roof based on your specific solar panel.
3. Check the required setbacks for rooftops. Generally, 3 feet is an acceptable setback.
4. Locate the roof's rafters to center the truss, which will be the supporting spot for the rails.
5. Install flashings (supporting structures that tighten the modules to the roof).

6. Place the rails.
7. Add grounding bolts and wire management clips.
8. Secure the modules to the mounting system by using clamps and T-bolts.

Solar panels mounted on rails
Source: FOEN solar bracket

Ground-Mount

For ground-mounting systems, the steps for the installation are as follows:

1. Excavate the ground to provide enough space for foundations. The type of foundation depends on the soil type in your area.
2. Concrete foundations or helical piles are installed.

3. The base of the mounting system is fixed to the foundations using bolts.
4. Vertical pipes are installed and fixed to the base of the mounting system.
5. Rails are installed and attached to the structure. Unistrut or Unirac are both great materials to work with.
6. Cross rails are installed if needed to provide additional support for the structure.
7. Middle and end clamps are used to adjust the solar panels to the rails.
8. A tilting mechanism can be installed for summer or winter.

Ground-mounted solar panels
Source: Wikimedia.org

RV's

RV solar panel systems have another installation process, similar to roof mount, but not equal. The process is briefly explained as follows:

1. Organize the available space on the roof and make a schematic of the distribution of panels on the rooftop.

2. Install the mounting brackets while the solar panels are still on the ground. This will save you additional work on the roof.
3. Create the setup process for the charge controller and batteries. From a roof perspective, figure out the shortest path for the wires from the roof to the charge controller's location.
4. Drill a hole through your roof to pass the wires.
5. Install a waterproof cable entry plate in the position of the hole to pass the wire output of the solar panels.
6. Drill the corresponding holes to secure every module according to the pre-drilled hole positions located in the mounting frame of every module.
7. Secure them to the roof by using bolts and screws together with silicone to make it waterproof.

There are many different types of mounts for placing solar panels on the roof of your RV. Some people prefer not to drill in their roof and use brackets to stick the panels to it.

Mounting solar panels on RV roof with brackets
Source: JdFinley.com

Items used for mounting:

- 90° brackets from your hardware store.
- 3M 5200 adhesive (no sealant) or VHB tape.

If you have selected flexible, thin solar panels, the installation is easier since you can use strong double-sided tape to adhere the panels to the roof.

All vehicles are designed to be aerodynamic, which means reducing air friction while traveling. If the vehicle is not streamlined, noise will be created, which you will hear when driving it at high speeds. Therefore, try to reduce air friction as much as possible.

Because of the air friction, the solar panels could come loose, which you do not want to happen while driving. Therefore, linking all your panels with a steel cable is recommended. Then, if one panel comes loose, it won't fly away and damage something or hurt someone. Also, periodically check your roof for any loose mounts.

Drill-free solar panel mounts

Sailboats

The procedure for sailboats is similar to that for RVs. Most of the time, you will need to use a boat Bimini cover to place the modules. When you have sized your Bimini cover, you will need to find similar suction cups or plastic attachments that can go through the holes of the panels and lock them in place.

A popular method is to stitch a flexible solar panel to the Bimini cover. Instead of using the suction cups at the corners, you install a special cloth that you can stitch to the bimini. Be careful that it doesn't shade the panel itself.

I recommend wiring in parallel because the boat will have quite some shade. Keep in mind that the solar panel's voltage needs to be 5V higher than the battery's. Therefore, you can be limited to a 12V battery.

Installing solar panels on a Bimini
Source: www.sailrite.com

Cleaning Panels

The Effect of Soiling

Maintenance is an important factor to consider when installing solar panels. Dirt or dust on the panels' surfaces can severely affect their performance. With regular maintenance procedures, this PV energy loss (commonly called soiling) should not be higher than 2% of the annual energy yields.

Extreme cases with considerable dust and no cleaning can reduce the modules' performance by up to 30%.

The following image shows a graph with the differences in the I-V curve and the MPP before and after the cleaning procedure for a small PV system. As you can see, the power output is reduced from 3kW to 2.15kW, which is a significant power output reduction and why it is so important to clean your solar panels regularly.

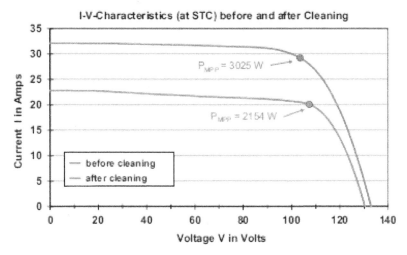

I-V curve after and before cleaning
Source: Haberlin Heinrich.
"Photovoltaics System Design and Practice"

How to Clean Your Solar Panels?

The solar panel cleaning procedure is straightforward. The first step is to disconnect the PV system by turning off the solar disconnect switch. This is for safety concerns.

Then, you need to get a sponge and submerge it in preferably de-ionized water. Avoid regular cleaning water since the minerals inside will adhere to the module's glass.

Avoid using cold or hot water to clean the modules. High-temperature differences are not good for them. Using the sponge, start cleaning each module until any noticeable dirt is removed.

Soft brushes can also be used. Telescopic cleaning poles will make cleaning your modules easier if they are located in inaccessible spots. Another option is to use a low-pressure hose to clean the modules. Pressure should always be under 580psi (40bar).

Never use laundry detergents, bleach, or any other product. All that needs to be used is distilled water and, if desired, a little bit of dishwashing soap. You can also use specially designed solar cleaning equipment to make the task easier.

When to Clean Your Solar Panels?

The time to clean solar panels is important. At noon, solar panels are producing electricity at peak performance. Choosing to do maintenance during this period is not good for two reasons. First, if you clean at this time, you will lose the most important energy yields of the day, and second, at noon, solar panels are hot, making them difficult to clean.

Solar panels can reach temperatures of over 158°F (70°C). The ideal time for cleaning is early morning or late afternoon, as solar panels' performance is lower and they are cooler at these times. Reducing the chances of creating micro cracks because of the sudden temperature difference between the hot solar panel and the cold water.

Solar Panel Lifespan

The lifespan of a solar panel depends on the brand and model type. However, currently, most rigid solar panels feature a 25-year life warranty.

It is important to notice that this is not the same as the product warranty (which is generally close to 10 years), but it is an estimated life expectancy of solar panel production. The warranty performance states how long the solar panels are expected to work and produce at least 80% of their power output in the initial year.

The best solar panels available can reach almost 30 years of performance and have a 15-year product warranty. The life warranty performance is in the module's datasheet.

Panel Voltage

As discussed earlier, current flows from areas of higher to lower voltage. Solar panels, designed to be power sources, must have a higher voltage than the load they're powering to drive this current flow.

The solar panel voltage must be high enough to charge the battery bank. For example, to charge a 12V battery, a solar panel with a voltage output of at least 17V is required.

For a 24V battery bank, the solar array should produce at least 29V. Some larger solar panels can already output 30-40V, directly charging 24V batteries without additional panels in series. To ensure proper charging, the solar panel's maximum power voltage (Vmp) should ideally be about 5V higher than the battery's voltage.

If additional voltage is needed, you can wire multiple panels in series until the voltage is sufficient, as long as it remains within the limits of your MPPT charge controller.

Buying Used Solar Panels

Solar panel installers will always provide you with new PV modules. However, occasionally, you will find available modules on the web set for sale at lower prices, but they are used.

Purchasing used solar panels for residential or commercial installations presents multiple problems that far outweigh the possible economic benefits.

Used solar panels can be on sale for different purposes, but mainly there are two reasons:

- First, the homeowner no longer wants to use all the panels for his/her PV system and wants some money back.
- Second, the solar panels suffered some damage, or their performance was not what was expected from them.

This can be a serious problem if you choose to purchase multiple used solar panels since you can end up investing a substantial amount of money and not receiving the expected performance. Solar panels also degrade over time, and it is hard to precisely estimate the number of years that the solar panel has been working.

Used solar panels have another problem related to technology. The solar panel industry has quickly evolved in the last decade. Solar panels from ten years ago are different in terms of power output, technology, and efficiency from solar panels today.

The only recommended case for purchasing used solar panels is through companies specializing in selling them used panels, such as Santan Solar, based in the U.S.

Now that we have discussed batteries and solar panels, let's examine a component that connects these two: the solar charge controller.

Charge Controller

What is the Task of a Charge Controller?

A charge controller is an electronic device designed to regulate the rate at which electric current is added to a battery. It has several functions. The first one is to safely operate the battery's charging process to ensure its long lifespan.

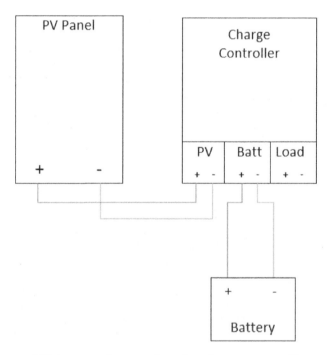

Wiring a solar panel to the charge controller

During the charging process, the charge controller measures the battery's state of charge (SOC). Based on the measured value, the charge controller increases or decreases the electric current to comply with particular battery charge stages.

The principal stages involved in the charging process for lead acid that every charge controller should comply with are:

- **Bulk:** high electric current charge increases the voltage of the cell.
- **Absorption:** stabilizing voltage requires slowly reducing the charging current.
- **Float:** the final process requires a very low charging current.

The following image shows a sample of the charging process in a lead-acid AGM battery using a charge controller.

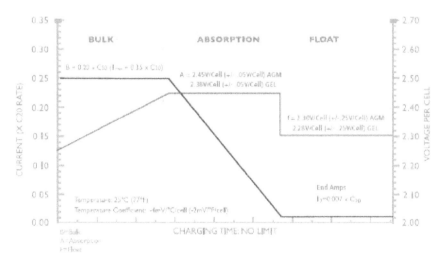

Charging cycle of a VRLA battery
Source: Rolls Battery Manual

The charge controller can handle several battery types, each with a different charging profile. Lithium, for example, does not have absorption and float; it only has bulk. Make sure the charge controller you buy is compatible with your battery.

The charge controller efficiently extracts solar power from the PV modules and adjusts the solar array's power output to the battery's required voltage (generally 12V, 24V, or 48V). By connecting solar panels in series and parallel, the string's voltage can reach higher than the battery bank's nominal value.

Another function is that the battery's voltage is higher during the nighttime than the solar array, which could make the energy flow from the batteries to the panels. Therefore, an important duty of the charge controller is to block the discharge at night to avoid these reverse currents.

If the solar panels were to be connected to the battery directly without a charge controller, you would be overcharging the battery, which leads to damage and eventually destroys it. The charge controller frees the system designer from closely matching the PV voltage to the battery's voltage and allows them to set longer strings than they could have been without the charge controller. Never connect a solar panel directly to a battery.

Different Charge Controllers

According to their operation modes, charge controllers can be divided into two main groups: PWM and MPPT charge controllers.

PWM

The Pulse-Width Modulated (PWM) charge controller was the first model to appear on the market and is the most basic (and cheapest) type of controller.

The PWM charge controller acts like a switch that connects the output of the photovoltaic modules with the battery bank. Once the switch is closed, the voltage in the charge controller's terminals will be the battery's nominal voltage.

The charging process of the PWM model consists of closing the switch during the first stage to the maximum possible current value as the voltage gradually increases. When the voltage reaches the absorption voltage value, the current decreases slowly by disconnecting and reconnecting the switch multiple times.

This creates a pattern of small pulses until the current drops to zero. The following image shows the charging process of a PWM charge controller for a lead-acid battery.

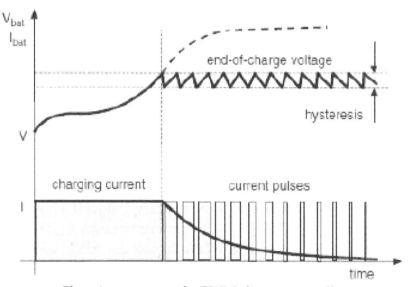

Charging process of a PWM charge controller
Source: A.Luque y S.Hegedus.
"Handbook of Photovoltaic Science and Engineering"

257

The PWM charge controller always sets the output voltage close to the nominal voltage of the battery bank (generally a little higher to be able to charge it). Then, the regulator will provide the corresponding current in the I/V curve of the solar array.

The following graph shows an example of the output of a PWM regulator for a 12V battery. As can be seen, the PWM controller does not operate at the maximum power point (81 Watts).

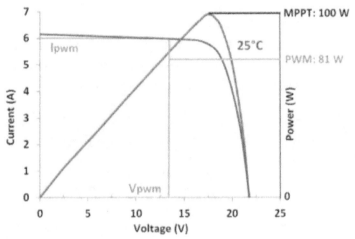

PWM and MPPT charge controller curve comparison
Source: Victron Energy

When using a PWM charge controller, you must closely match the solar panel's voltage to the battery bank. If you do not closely match the voltage to the battery voltage, there will be high losses. This means that when operating a PWM controller, you will likely have to wire the panels in parallel, which increases wiring cost.

Here is an example of what happens when you connect a PWM charge controller.

The voltage into the charge controller will be 20 Volts. The PWM charge controller cuts the voltage to 13 Volts to charge the battery, so you lose about 7 Volts because the PWM doesn't track the MPP (maximum power point).

The power loss over the PWM charge controller will be 35%, resulting in a 65% efficiency. Therefore, it's essential to match your solar panels to the voltage of your battery bank. PWM charge controllers are available in different voltages to match the battery at 12V, 24V or 48V.

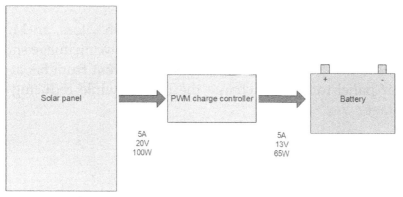

Loss of energy when using a PWM charge controller

MPPT

The second type of charge controller is the maximum power point tracker (MPPT). This controller is a DC-DC converter (a device that transforms a DC signal into another DC signal with other parameters).

This type of controller's operation mode involves adjusting the voltage in the output terminals according to the voltage required to charge the battery bank.

At the same time, the controller tracks the changes in the solar array across the day in the I/V curve and locates the maximum power point of the curve every time. After locating this point, the controller determines the amount of electric current needed to provide the same amount of power as the MPP would provide but at the battery's nominal voltage.

Instead of simply assigning the corresponding electric current in the I/V curve (like the PWM controller does), the MPPT model increases the electric current to reach the maximum power point.

Compared to a PWM, which can have a 35% loss, an MPPT charge controller has a 2 to 6% loss. The following image shows that the MPPT controller increases the current from 5A at the solar panel to 7.5A to charge the battery while reducing the voltage.

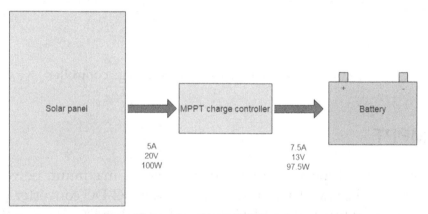

The efficiency of an MPPT charge controller

PWM or MPPT?

As previously explained, a PWM charge controller does not operate at its maximum power point. This means that the power output obtained from a solar panel array when using a PWM model is much lower than that obtained from the MPPT model.

In commercial-based or residential systems, the MPPT charge controller must always be the preferred choice since the extra energy obtained from the I/V curve exceeds the additional cost of the MPPT model.

A particularly important factor to consider when selecting the charge controller is the voltage. If the solar array has many solar panels in series, the voltage will be higher. As the voltage increases, the maximum power point (MPP) will be further from the nominal voltage of the battery bank.

If you have multiple solar panels wired in series, the PWM charge controller would have much more losses than the MPPT model. In this case, it is recommended to choose the MPPT charge controller. Having more solar panels wired in series also reduces the amount of current and, therefore, the cross-sectional area of the wires that would be needed, which translates into cost reductions.

There is no specific voltage at which you should consider switching from a PWM to an MPPT. Nevertheless, remember that you need a higher voltage to charge the batteries. Therefore, if you have a battery bank at 12V, you will need at least 17V at the source (panels). PWM charge controllers are usually limited to an input voltage of around 50 Volts.

Another factor to consider is shading. Under shading conditions, the MPPT charge controllers will perform much better than the PWM models because the MPPT models can track the array's maximum power point (which will be affected by the presence of shade). Therefore, if your solar panel installation is expected to have shading conditions, choosing an MPPT model is better.

Despite their advantages, MPPT charge controllers can be considerably more expensive than PWM models. Therefore, the decision comes down to balancing costs and performance. For a small system up to 200W of solar, using a PWM is acceptable. For larger systems, I recommend using an MPPT.

A PWM charge controller can become very handy for small loads, or a trickle charger to reduce costs.

Because off-grid applications like vans, boats, or cabins have limited space to put solar panels, we need to extract as much energy as possible from the panels. The best way to do this is with an MPPT charge controller.

Selecting a Charge Controller

When choosing a charge controller, you must consider multiple factors. We will look at a datasheet from a widely used charge controller, Renogy's Rover series.

Model	RVR-20	RVR-30	RVR-40
Nominal system voltage	12V/24V Auto Recognition		
Rated Battery Current	20A	30A	40A
Rated Load Current	20A	20A	20A
Max. Battery Voltage	32V		
Max Solar Input Voltage	100 VDC		
Max. Solar Input Power	12V @ 260W 24V @ 520W	12V @ 400W 24V @ 800W	12V @ 520W 24V @ 1040W
Self-Consumption	≤100mA @ 12V ≤58mA @ 24V		
Charge circuit voltage drop	≤ 0.26V		
Discharge circuit voltage drop	≤ 0.15V		
Temp. Compensation	-3mV/°C/2V (default)		

Renogy Rover series charge controller datasheet

Voltage of the Battery Bank

One of the first things to consider when selecting a charge controller is verifying that its nominal output matches the battery. Typical applications are 12, 24, or 48 Volts.

Rated Current

This is the value displayed on the charge controller. It is one of the most important features when choosing your charge controller. This number indicates the amount of current that will go to the battery at the specified system voltage.

Let's explain this with an example. We have two solar panels with the following specifications:

MODEL: SPR-E-Flex-100

Rated Power (Pmax)[1] (+/–3%)	100	W
Voltage (Vmp)	17.5	V
Current (Imp)	5.80	A
Open-Circuit Voltage (Voc)	21.0	V
Short-Circuit Current (Isc)	6.20	A
Maximum Series Fuse	15	A

Standard Test Conditions: 1000 W/m², AM 1.5, 25° C
Suitable for ungrounded, positive, or negative grounded DC systems
Field Wiring: Cu wiring only, min. 12 AWG/4 mm², insulated for 90° C min.

If we wire both panels in series, the voltage will double while the current stays the same. So we get 35 Volts and 5.8 Amps.

However, we do not charge our batteries at 35 Volts. The charge controller steps down the voltage to charge the battery. This conversion increases the current (only with an MPPT charge controller).

The combined power of both panels is 200 Watts. Let's calculate the current at the nominal battery voltage of lithium which is 12.8V (12V for lead-acid).

$$Charging\ current = \frac{200\ Watts}{12.8\ Volts} = 15.6\ Amps$$

You can see that the battery charging current is higher than the previously calculated 5.8 Amps. This means that you would need a 20-amp charge controller to fully benefit from it.

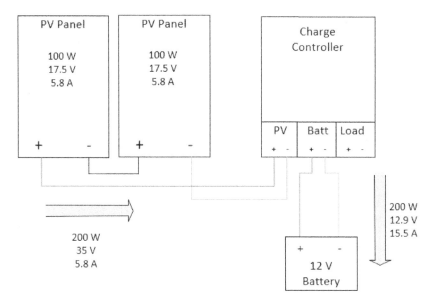

Power conversion of an MPPT controller and 12V battery

You can also undersize the charge controller by using a 15A charge controller. The 0.6A will get 'lost' because the max output current of the charge controller will be 15A. This won't damage the charge controller. However, running a charge controller at it's maximum output can reduce the lifespan because it will get warm. And heat is not good for the longevity of electronics.

Let's repeat the calculation for a 24V battery.

$$Charging\ current = \frac{200\ Watts}{25.6\ Volts} = 7.81\ Amps$$

As you can see, the charging current is reduced because we are charging at a higher battery voltage. When the charging voltage increases, the charging current will decrease. That's why it's cheaper to use a higher voltage battery bank. You will also save on wiring costs because you need thinner wires with lower current.

Charging at a higher voltage will increase the charge controller's efficiency because the charge controller doesn't need to step down the voltage to 12 Volts. We would now need a 10 Amp charge controller instead of a 20 Amp charge controller.

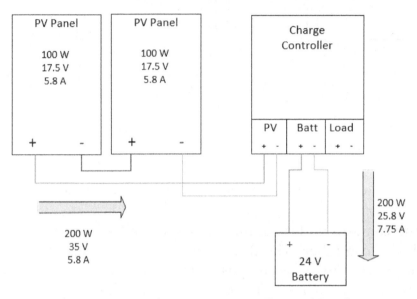

Power conversion of an MPPT controller and 24V battery

Increasing the battery bank's voltage allows you to use a smaller charge controller because the charging current will be lower. As a result, the charge controller with a lower charge current will be cheaper.

Load Current

As mentioned before, we generally do not use the load terminals of the charge controller unless it's a small load. It is best to use the battery terminals or busbar to power AC and DC loads combined with a low voltage disconnect (if you have lead-acid). In the case of a lithium battery, the BMS will shut down the battery if a certain low voltage is met (usually 10V for a 12V LiFePO4 battery).

Input Voltage

Another factor that must be considered is the input voltage.

As previously discussed, the solar panel voltage varies across the day, depending on the temperature. Moreover, it also changes depending on the number of solar panels connected in series. Therefore, when sizing a solar array, you need to estimate the number of solar panels that need to be connected in series. The more panels in series, the higher the voltage.

The charge controller that you selected will have both a maximum and a minimum input voltage from the solar array that it can handle. Whenever choosing the charge controller, you need to make sure that the input voltage does not go over that range, otherwise the charge controller will get damaged.

You must consider the effect of temperature on the cell voltage, as explained in the solar panel section. In other words, the maximum input voltage that the charge controller can accept should be associated with the voltage of the string under minimum temperature conditions.

Meanwhile, the minimum input voltage that the charge controller should accept is related to the voltage of the string under the maximum temperature conditions. Most small MPPT charge controllers have a maximum input voltage of 100 or 150VDC.

Lets say you have to following solar panel:

MODEL: SPR-E-Flex-100		
Rated Power (Pmax)[1] (+/–3%)	100	W
Voltage (Vmp)	17.5	V
Current (Imp)	5.80	A
Open-Circuit Voltage (Voc)	21.0	V
Short-Circuit Current (Isc)	6.20	A
Maximum Series Fuse	15	A

Standard Test Conditions: 1000 W/m², AM 1.5, 25° C
Suitable for ungrounded, positive, or negative grounded DC systems
Field Wiring: Cu wiring only, min. 12 AWG/4 mm², insulated for 90° C min.

CE

RoHS

We can calculate how man panels we can wire in series for a 100V max input voltage. The 1.25 is a safety factor for cold weather. When it's freezing outside and the sun shines, the voltage can be higher than what the panel is rated for.

$$\frac{Max\ input\ voltage}{Voc \times 1.25} = \frac{100V}{21V \times 1.25} = \frac{100V}{26,25V} = 3.8\ panels$$

We can have a maximum of 3 panels in series for a 100V charge controller. If you wire 4 panels in series, the charge controller can get damaged on a cold sunny day.

If the voltage exceeds the maximum input of the charge controller, the charge controller will get damaged.

Power

Another factor is power. The charge controller is capable of handling a specific amount of power input from the solar panel(s).

As you already know, power is the product of voltage and current. The power of a solar string or array is the same in every configuration (series or parallel). The voltage refers to the nominal battery voltage.

We can see that a higher battery voltage allows a larger solar panel input. This is because the components of a charge controller are selected according to current. Raising the battery voltage by a factor of 2 (from 12V to 24V) will decrease the current by a factor of 2. That's why we can have a higher power input when the battery voltage is higher.

| Max. Solar Input Power | 12V @ 260W |
| | 24V @ 520W |

Maximum input power is related to battery voltage

Efficiency

Efficiency is important to consider. When deciding between two charge controllers, the model with higher efficiency is preferable. MPPT controllers are the best option.

Since charge controllers are electronic devices intended to act as intermediaries between the charging source and the load, then charge controllers should transfer as much energy as possible.

Therefore, efficiency should always be located above 94%. Efficiency will be higher (98%) when you choose a 24V or 48V battery system compared to a 12V battery system.

Operating Temperature Range

Each charge controller has a recommended operating temperature range, so it's important to place it where these limits won't be exceeded. Make sure there's adequate space around the controller for ventilation, and avoid installing it in areas that may become wet.

Temperature Compensation

As explained in the battery section, temperature changes can severely affect batteries. Battery capacity can drastically reduce as it approaches 32°F (0°C) and even more if it lowers. Combining this temperature effect with high load demands can lead to a severely reduced state of charge (SOC) in your battery.

Voltage charging setpoints are usually established for charge controllers assuming 77°F (25°C), which is the standard for batteries. However, as the temperature increases, the charging voltage lowers. If the temperature decreases, the charging voltage rises. The charge controller must integrate temperature compensation to avoid overcharged (which can lead to gas expulsion) or undercharged (leading to sulfation in lead-acid) batteries.

Chargers can either have a fixed temperature compensation voltage (e.g., 5mV/°C/cell) or an adjustable setpoint. An adjustable setpoint is advantageous since battery manufacturers can recommend different temperature compensation values.

For instance, Rolls and Victron Energy Batteries recommend 4mV/°C/cell while Crown and Deka recommend 3mV/°C/cell.

The charge controller's temperature compensation consists of using the nominal system voltage, the nominal charge voltage at 77°F (25°C), the temperature compensation rate, and the battery temperature.

As a reference example, let's assume a 24V system, with a charge voltage of 28.6V, along with a temperature compensation rate of -5mV/°C/cell and a battery temperature of 113°F (45°C).

Since the system is 24V lead-acid, it has 12 battery cells (2V per cell). Then,

$$Temp_{comp} =$$
$$[Temp_{rate} \times Numb_{cells}] \times [Batt_{temp} - 25°C] + Charge_{voltage}$$

$$Temp_{comp} =$$
$$\left[-0,005\ V\frac{°C}{cell} \times 12cells\right] \times [45°C - 25°C] + 28.6V$$

$$Temp_{comp} = -0.06V \times 20°C + 28.6V = 27.4V$$

The new charging voltage for that battery bank would be 27.4V instead of 28.6V. As you can see, the voltage is reduced.

There are two types of charge controllers in this aspect. Some models already integrate an internal temperature sensor and can perform this function directly. These models are quite helpful when the battery bank and the charge controllers are located inside the same enclosure and have similar temperatures.

There may be occasions in which the battery and the charge controller are located in different enclosures. In these cases, it is appropriate to choose a remote temperature sensor. Remote temperature sensors consist of a wire with a terminal pin to be connected to the charge controller and a sensitive probe to be used for testing and measuring temperature. This is also recommended when the temperature changes across the year step over 5°C.

Most of the higher-end MPPT charge controllers have a port where you can add a temperature sensor. The charge controller will use the temperature and calculate the ideal charging voltage according to your battery type. Then, you put the sensor in the same compartment as your batteries. You can use tape to stick the sensor to the side of your batteries for the most accurate compensation.

If you are using lithium batteries, you should use a temperature sensor located on the BMS. If you charge the battery when it's freezing, it's permanently damaged. The temperature sensor can detect this and disable the battery's charging. Some lithium batteries have low-temperature protection, meaning that they will stop charging if the battery goes below freezing. Discharging is still okay below freezing. It will resume charging whenever the battery is above freezing again.

Temperature compensation sensor

Temperature compensation on the Renogy charge controller

Connecting the Charge Controller

The procedure of wiring a charge controller into the system is very simple.

Charge controllers generally have two output terminal sets: one for the battery and the other for the load. They also have a single input terminal set connected to the solar panels.

When wiring the charge controller, connect the batteries to their corresponding output terminals. Then, you connect the loads to the load outputs (if any). Then, wire the solar panels to the charge controller.

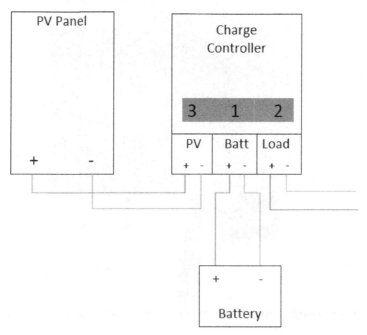

Connection sequence for the charge controller

The load terminal on the charge controller can only supply a limited amount of current, so I don't recommend using it. It's better to connect the DC fuse box straight to the battery or busbar.

Remember that the charge controller input and output will have a maximum gauge size connected to the terminals, which will depend on the charge controller's electrical current output.

If you want to do maintenance and disconnect the PV system, always disconnect the DC switch between the charge controller and the solar panels. Once this is done, remove the load connections before removing the battery's connections from the charge controller.

In other words, solar panels are the last ones to connect and the first ones to disconnect.

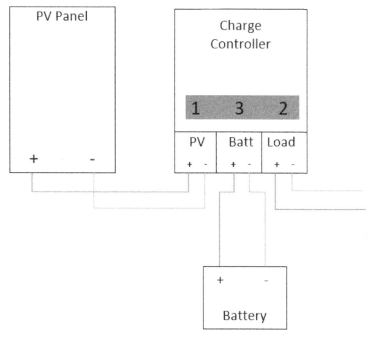

Disconnecting sequence

Always check the polarity of the wires with a multimeter before connecting them to your charge controller. Ensure the solar power input voltage is not higher than the allowable max input voltage.

Multiple Charge Controllers

There are many reasons why you would want multiple charge controllers. These are:

- Adding more solar panels to your system.
- Adding a panel with a different specification than the one you already have.
- Separate panels from each other because they receive shade at different times of the day. For example, on a boat or multiple sides of the roof.

Adding more charge controllers to the same battery isn't a problem. The charge controllers detect the battery's internal resistance to determine the charging current. If the internal resistance is low, the battery is empty. If the internal resistance is high, the battery is full. Both (or more) charge controllers will charge their total current into the battery bank.

You have to make sure that the battery can handle the combined current of both charge controllers. This will be easier with lithium because it can tolerate a higher charging current.

If you have lead-acid batteries, you must be careful not to charge lead-acid too quickly with too much current. Read the manual on the maximum recommended charging current or C-rate.

You also need to disable the equalizing function on the other charge controllers and only enable it on one controller (the master). Otherwise, every charge controller will randomly equalize the lead-acid battery.

The charge controllers do not need to be able to communicate with each other. Some charge controllers can connect through a cable or Bluetooth. This is a nice feature, but it is unnecessary to operate multiple charge controllers for one battery bank safely.

The following image shows how to wire multiple charge controllers to one battery bank. It is recommended that the wires connecting the charge controller to the battery be the same length. This will make the wires' resistance the same, allowing them to charge equally.

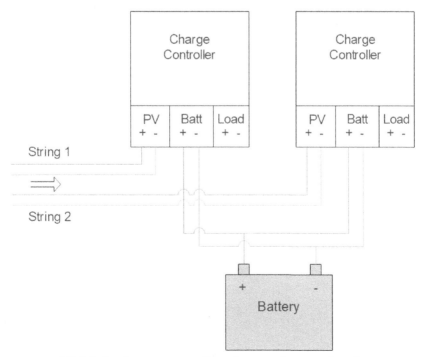

Multiple charge controllers for one battery bank

Lithium batteries can now charge at 1C rate. This means that a 12V 100Ah battery can get charged at the following current:

$$100Ah \times 1C = 100A$$

However, when you have a lead-acid battery, it's recommended C-rate is 0.2C. That means you should only charge a 12V 100Ah battery at the following current:

$$100Ah \times 0.2C = 20A$$

Let's look at the last component of the solar power system. The Inverter.

We can see that lithium is better suited for high-current charging and discharging.

Inverters

What is the Task of an Inverter?

Finally, the inverter is the PV system's last but essential component.

The inverter's most important task is to convert the DC signal in the batteries into the AC signal consumed by most loads. As explained at the beginning of this book, there is an important difference between DC and AC signals.

Since solar panels generate DC, and most household loads require AC signals, the inverter acts as a DC-AC converter that can or cannot synchronize (depending on the model) with the signal obtained from the power grid (grid-tie inverters).

However, there are different types of inverters, and depending on the type, they may have alternative functions. Larger inverters of 1,000 Watts or more have large capacitors to create the AC signal. The capacitors will charge immediately when you connect the inverter to a battery, creating sparks. Sparks are not desired because they can damage the electronics and terminals.

That's why it's best to use a resistor in series to limit the current flow into the inverter on the initial connection. These resistors are called pre-charge resistors. They range from 5 ohms to 30 ohms.

Connect the negative lead from the inverter to the battery first, and then use the pre-charge resistor to touch the positive wire from the batteries with the positive terminal of the inverter for a few seconds. This will eliminate sparks, and the capacitors in the inverter will charge.

25-Watt, 30 Ohm resistor

Types of Inverters

Off-grid

The most common type of inverter for applications in RVs, cabins, and boats is the battery-based inverter (also applicable for inverter/chargers) suitable for stand-alone setups.

This type of inverter works independently from the power grid and can be used with lead-acid or lithium batteries. It will be connected directly to the batteries and will have a DC input value at the battery's nominal voltage (generally 12V, 24V, or 48V). This type of inverter will create its own AC signal.

Always remember that the AC loads must be connected to the AC side of the inverter, which is in the output terminal. AC loads must never be connected to the charge controller because the charge controller (no matter what type) will always work in DC. The wiring for this type of inverter can be found in the following image.

Never connect the AC out of an off-grid inverter to the grid. This will damage the inverter.

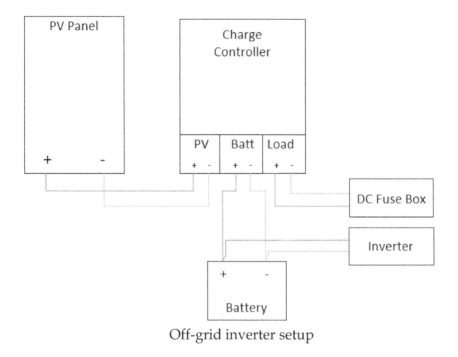

Off-grid inverter setup

Grid-Tied

The most common type of inverter available is the grid-tied model, which is used when the power grid is available for connection. Another name for this inverter is a string inverter.

This applies mainly to residential, commercial, and utility-scale applications. This type of inverter needs a signal from the power grid to synchronize since it cannot create its own AC signal. Therefore, when the power grid goes down, the inverter shuts down. This protects people working on fixing the electrical lines from being electrocuted.

The input of this inverter is the solar panels. It can perform maximum power point tracking, while the output is an AC source signal at 50 or 60Hz at 120 or 230V. The output signal of the inverter will have a slightly higher voltage than the grid. This is to make the electricity flow from the inverter to your home appliances instead of power from the grid. If the voltage level in the grid rises, the inverter will track the change and increase its output voltage level.

A grid-tied inverter will not work for you if you have an RV, off-grid cabin, or boat. If you are using grid-tied, you need permission from your electrical company to feed electricity back into the grid. The following image shows a schematic of the connection of a grid-tied inverter.

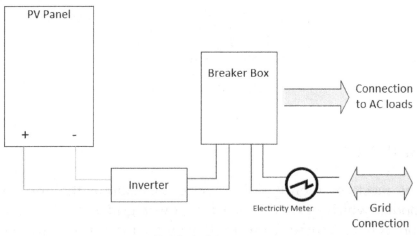

Grid-tied inverter

Hybrid-Inverter (all in one)

The hybrid inverter is the most advanced and complex inverter available. This device integrates the benefits of an MPPT charge controller, inverter, and battery charger into a single product. An automatic transfer switch could also be included.

The hybrid inverter can receive the signal from the solar panels and track the maximum power point to extract solar energy, just like an MPPT charge controller.

It can use this energy to store it in a battery, combining the advantages of available power from the grid with energy storage. Some models (called grid-tied hybrid inverters) can synchronize with the grid when available. They can also generate their own AC signal when a blackout occurs.

Then you have other hybrid inverters that can take energy from the grid to recharge the battery but cannot deliver power back to the grid (because they do not synchronize with the grid). These are called off-grid hybrid inverters.

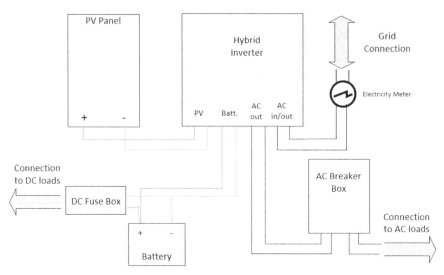

Hybrid-inverter schematic

One advantage of hybrid inverters is that they can charge the batteries without the need for a charge controller through the integrated battery charger from the grid.

Technically, you could buy the inverter for that purpose. However, you would be underestimating its potential as you still constantly depend on shore power to charge the batteries and power your loads. By installing solar panels, you could move with the RV and travel for days without worrying about the availability of shore power.

The hybrid inverter will use the battery bank to make an AC signal if there is a blackout or if you are on the road. If it detects that your solar panels or onshore power (at a campground) are producing power, it will use that power to charge the batteries.

The disadvantage of a hybrid inverter is that it tends to consume quite some power in standby mode. If you only use DC, the inverter will be idling, constantly consuming power. A 48V 3kW hybrid inverter might consume 30W. So for 24h, this becomes (30W*24h=720Wh), which is quite significant. For example, a single 12V 100Ah lithium battery can hold 1280Wh. Some hybrid inverters can turn off the inverter with a separate switch or power-saving mode. A popular off-grid hybrid inverter from EG4, called the 6000XP will consume 50W of idle power consumption. That's equivalent to 1,200Wh or a single 12V 100Ah battery in one day.

The hybrid inverter can be used in the following case: You are plugged in and draw 20 amps from shore power, the solar panels produce 20 amps, and your AC unit draws 40 amps. Now, you can run your AC without drawing power from the battery.

Be careful about cheap hybrid inverters. If the price difference is significant compared to others, they might use a PWM charge controller. Always check the specifications of the product carefully.

Some popular brands for hybrid or all-in-one inverters are EG4, MPP Solar ('maximum_solar' on eBay), Victron, and Growatt. Because of their fans, hybrid inverters can make a lot of noise. Consider this when buying your hybrid inverter. Another point to watch out for is whether they include a grounding relay. We will talk about this in the chapter about grounding.

Inverter/Charger

This might be the best solution if you want a separate charge controller. This inverter converts DC voltage from the battery to AC loads and can also charge the battery itself. It's a cost-efficient solution to charge your batteries with shore power while on the road. An example of an inverter/charger is the Victron Multiplus.

Unlike the hybrid inverter, the inverter/charger does not include the solar charge controller. How to wire the inverter/charger can be seen in the following diagram:

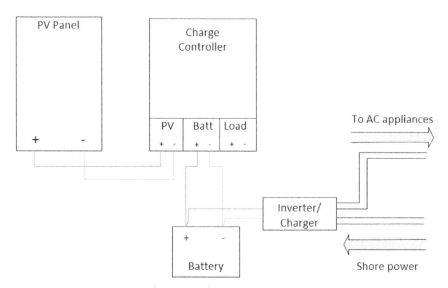

Wiring diagram for an inverter charger

Apart from using shore power, you can also use a portable generator to charge the batteries by connecting it to the inverter/charger's AC input. A manual switch can be added to switch from shore to generator power on the AC input.

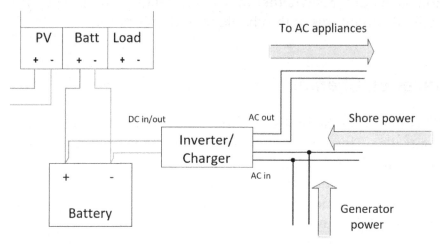

Wiring inverter charger with shore power and a generator

Inverters Output Signal

Inverters can also be classified by the shape of the signal wave at their output and their configuration type.

Square Wave

These are considered the oldest technology available for inverters. Their working mechanism tries to resemble the periodic shape of an AC signal by flipping the voltage directly from negative to positive and back again.

Unlike a typical AC signal, there is no pause through zero between the negative and positive voltages. Due to this alternating pattern, they resemble the shape of a square.

Today, square wave inverters are not considered for many applications because most loads would heat up when using them. Electronics require smoother signals free of harmonics (noise components).

Square wave inverters have a much higher noise level than sine wave inverters, which creates a real disturbance. Despite these flaws, square wave inverters can be much more affordable than their sine wave counterparts. This is because the square wave is an older, less complex type of inverter.

Square wave inverters are no longer available for purchase. They have been replaced by modified sine wave and pure sine wave inverters, which offer improved efficiency and compatibility with modern electronics.

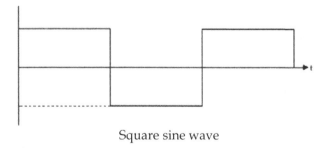

Square sine wave

Modified Sine Wave

The modified sine wave inverter is another version of the inverter that came after the square wave type. It has better performance and is more similar to an actual AC sine wave signal.

Unlike the square wave type, the modified sine wave adjusted the ups and downs of the output signal to match the shape of an actual AC sine wave signal as closely as possible.

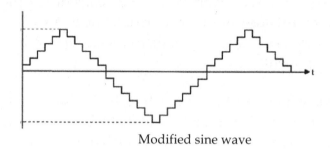

Modified sine wave

Despite having better performance than the square shape type inverter, the modified square wave inverter will still have a noticeable harmonic distortion that will not allow it to run sensitive electronics such as laptops. Besides, you will likely hear a "buzz" when operating it. Modified sine wave inverters are cheaper than pure sine wave inverters.

Pure Sine Wave

The pure sine wave inverter is the most efficient, technological, and undoubtedly ideal option for any load. Its signal resembles exactly the one that could be measured in your house's power outlet, meaning a perfect sine wave.

Sensitive electronics need the cleanest signal available to work without any risk of damage. Despite that, they will be more expensive. They will not be as noisy as the square wave model, they will not heat up your appliances, and they will work better in cases where there is constant use across the day.

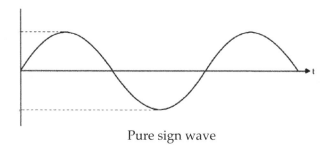

Pure sign wave

Utility Interactive for on-grid Connections

Inverters connected to the grid (grid-tied or hybrid inverters) have additional features that off-grid-based models do not need. Due to their connection with the grid, inverters can operate in different interactive modes depending on the conditions of the power grid.

The first type of connection available is the simple grid-tied mode, which exports all the energy generated by the solar panels back to the grid. Depending on your location, multiple compensation schemes are available.

For U.S.-based cases, schemes such as Net Metering and Feed-in tariffs compensate you for every kWh injected back into the grid.

Nowadays, inverters are capable of more. Some, known as smart inverters, can interact with the power utility operator to guarantee the safest operation of the power grid.

Using high-tech communication systems, a network operator can shut down all inverters within a perimeter if the solar power injection to the grid exceeds the grid's stability limits.

Network operators can also curtail or adjust the power factor of smart inverters to reduce or increase solar power injection into the grid. Some networks, such as the Hawaiian power grid, already operate under these mechanisms. Hybrid inverters have other functionalities as well. Some of them are:

- **Backup:** Isolating operation from the grid when there is a blackout.

- **Grid Zero**: This system does not allow selling energy to the utility. The inverter tries to avoid energy consumption from the grid, using only renewable energy sources and batteries.

- **UPS:** This is intended to increase the response speed to support specific loads (critical for data centers or your computer).

- **Mini-Grid:** Operates as an off-grid system, using the utility grid as a backup when the batteries are low.

- **Time-of-Use:** The inverter uses low-cost electricity rate periods to consume power from the grid and high-cost electricity rate periods to sell power back if possible.

Efficiency

Like any other conversion device, an inverter's efficiency is essential to ensuring the proper use of solar power.

Efficiency is the amount of power in the DC signal that enters the inverter compared with the amount of power in the AC signal that comes out of the device. To reduce energy losses, inverters should generally have efficiency values above 90%. However, the latest inverters will have efficiency values above 95%.

Automatic Load Shedding

Load shedding is a technical term for reducing the amount of load (power) connected to a generator, energy source, or electrical system. This ensures that the electrical system can uninterruptedly power to the most important (critical) loads.

In an off-grid solar power system, it is important to consider that during low solar radiation days, the availability of power may be lower than expected. That's when an automatic load-shedding control mechanism or device becomes handy.

Some inverters, like the SMA Sunny Island inverter and the Victron Multiplus with its essential load output, have already integrated this feature. However, you may also find separate devices that can perform this task. This will allow you to prioritize the loads you want to keep running in increased load demands or low electricity generation.

For example, when your battery gets low, you don't want to run your electric water heater or run the AC. An inverter can shut down an AC output when the battery reaches a specific SOC, or you can set an alarm once the battery goes below a certain charge and decide not to use high-power appliances.

Low Voltage Disconnect

Some inverters have a built-in low-voltage disconnect feature. When the batteries go below a specific voltage, the inverter automatically turns off the loads. This feature can be helpful if you do not use a separate low-voltage disconnect.

If you do not know about this feature, it might be frustrating to see that your inverter is shutting down when the battery is almost depleted. This can happen when you have a large 3,000W inverter connected to a single 12V 100Ah lead-acid battery. The battery cannot handle the large load, and its voltage will drop significantly, triggering the inverter's low-voltage disconnect.

Peak Power

Peak power is the maximum power that the inverter will provide in a short time. Surge power follows a similar concept, although with a shorter time frame. It is important to mention that this does not refer to the inverter's power capacity when sizing the system. The peak or surge power can be used as a reference for loads requiring extra power, such as motors, during the starting process.

Depending on the inverter model and brand, you will find some products that offer different peak power capacities under different time frames. For instance, a 2,000W inverter may provide a peak power of 2,500W over 30 minutes (just in case the load demand slightly increases). It may even provide a surge power of 4,500W for 5 seconds (only suitable for starting loads with motors like a fridge or A/C unit).

Low and High Frequency Inverters

Low-frequency inverters use large transformers, making them heavier and bulkier than high-frequency models. For instance, a typical 12V 1000W low-frequency inverter can weigh around 35 lbs (16 kg) due to its substantial transformer, often made from copper coils. This design gives low-frequency inverters several advantages, especially for systems that must handle high starting surges from inductive loads like air conditioners, refrigerators, and motors. These inverters excel in surge handling, allowing them to provide power stability in demanding applications.

In terms of lifespan, low-frequency inverters are generally more durable. Since they rely more on mechanical components than on delicate electronics, they can withstand larger surges and maintain functionality over time. This durability, however, comes at a price. Due to the materials used, particularly the heavy copper transformer, low-frequency inverters are usually more expensive than their high-frequency counterparts.

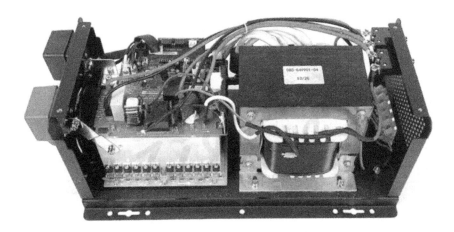

Low frequency inverter

High-frequency inverters, in contrast, are built using lightweight electronic components, including MOSFETs, which switch power at high frequencies without the need for large transformers. This design reduces the overall weight of the inverter significantly; for instance, a 12V 1000W high-frequency inverter might weigh only around 6 lbs (2.7 kg). This lighter build makes them easier to install and suitable for mobile setups like RVs or boats, where minimizing weight is important.

High-frequency inverters are also known for their lower standby or idle consumption, which can improve efficiency in setups with intermittent power needs. However, they're generally best suited for resistive loads — such as lighting or appliances without motors or compressors — since frequent high surges can reduce their lifespan. The reliance on electronic components like capacitors and MOSFETs can lead to faster wear compared to low-frequency models, especially in systems with demanding power spikes. Cost-wise, high-frequency inverters are typically more affordable because they use less copper.

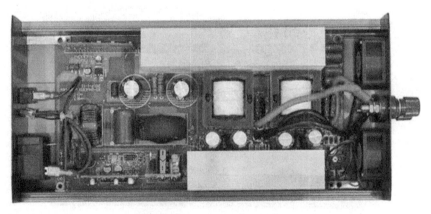

High frequency inverter

The choice between low- and high-frequency inverters depends on your specific needs. Low-frequency inverters are generally preferred for off-grid systems or setups where inductive loads and reliability are essential. High-frequency inverters, however, can be ideal for lighter applications with mostly resistive loads or for mobile setups where space and weight are primary concerns.

Some brands offer a hybrid approach, such as Victron, which combines low idle power consumption with both frequency types in their "Hybrid" models. Aims primarily uses low-frequency designs, while brands like Growatt often utilize high-frequency technology, though they offer some low-frequency models. If in doubt, weight is usually a reliable indicator: low-frequency inverters are significantly heavier due to their transformers, while high-frequency inverters are lighter.

In general, for applications requiring durability, especially in off-grid systems with variable loads, low-frequency inverters are a solid choice.

Parallel Inverters

It is possible to wire two or more inverters in parallel to increase the power output. However, these inverters need to sync their AC output with each other so their sine waves are in phase. One 3,000-watt inverter wired in parallel with another 3,000-watt inverter will deliver a total of 6,000 watts.

Most hybrid inverters can communicate with each other through an ethernet cable. Always check that your inverter can perform parallel functioning. Most stand-alone inverters cannot. If you combine the output of a regular off-grid inverter with another, one or possibly both will get destroyed.

Charging from Alternator

Solar energy is entirely dependent on the weather conditions of your location. Multiple rainy days may prevent your solar panels from completely charging your batteries. Over time, you may consume more power than your system was designed for.

Either one of them, having an alternative energy source to charge your batteries, is always preferable.

Vehicles can charge batteries with an additional energy source: the car alternator. Just as the vehicle moves and charges your starting battery, it can also power an auxiliary battery. There are two ways of doing this.

Battery Isolator Charger

The first option is to use an isolator charger.

The purpose of an isolator charger is to connect your vehicle's alternator to the solar battery and charge it when the car is running. However, when the vehicle stops, it will drain your starting battery. Therefore, the isolator charger ensures you *isolate* the starting battery from the solar battery when the engine is not running. This is the lowest-cost option, and I recommend it for lead-acid batteries.

For lithium batteries, direct charging from the alternator can damage the alternator and battery. This is because the charging process has no current limitation. Therefore, it is not recommended to use an isolator charger.

A car alternator is not made to charge deeply discharged batteries. The current demand from the alternator to the solar battery will be too high for the alternator. This makes the alternator provide more amps than it can safely handle, leading to overheating and, eventually, the destruction of your car's alternator. This is especially true with lithium batteries because they have low internal resistance.

Battery-to-Battery Charger (B2B)

The second option uses a battery-to-battery or B2B device, also known as a DC-to-DC battery charger.

This device is located between the starting battery and the solar battery. Its purpose is to take the car starter battery and boost or reduce it to provide a stable voltage under a multi-stage charging profile. This ensures the solar batteries are safely charged while keeping the main starter battery full.

Renogy offers a B2B device that connects to your car's ignition. It only runs when the car engine is running. The switches on the side allow you to select several options for battery type and battery voltage.

Battery-to-Battery charger from Renogy

The difference between a B2B and an isolator charger is that the B2B can perform a multi-stage charging scheme (the most optimal charging method). It should also come with a lithium charging profile. Victron has B2B chargers as well.

However, charging a 48V battery from a 12V starter battery is expensive. So, if you want to charge the battery with an alternator, I recommend using a 12 or 24V battery system.

Regarding B2B chargers, I can recommend Renogy and Victron.

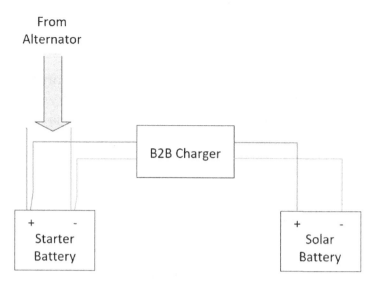

From
Alternator

B2B Charger

+   -
Starter
Battery

+   -
Solar
Battery

Wiring diagram for a B2B charger

When selecting a B2B charger, you must consider the maximum current. I recommend using 50% of the current your alternator is rated for. If your alternator is rated for 100A, I recommend a maximum of 50A to charge the battery.

Grounding

This topic is quite technical. If you prefer a video format, I have made a 3-video playlist on my YouTube channel where it can be digested easier. You can watch the playlist here: https://cleversolarpower.com/book/3videos

This playlist consists of three videos:

- Grounding
- Neutral ground bonding
- GFCI

Let's discuss grounding a stationary off-grid system, like one you'd set up for a cabin or home. Several components, such as your solar panels, inverter, and charge controller, need grounding.

There are two types of grounding we need to talk about:

- DC ground
- AC ground

These should stay separate because mixing them can cause problems.

If you measure the DC grounding on your inverter and the AC grounding, you will see they are connected to each other. But if you have an inverter/charger with a built-in grounding relay, this will cause grounding loops.

I like to keep things consistent, so I recommend separating your AC and DC grounding busbars.

Stationary Systems Grounding

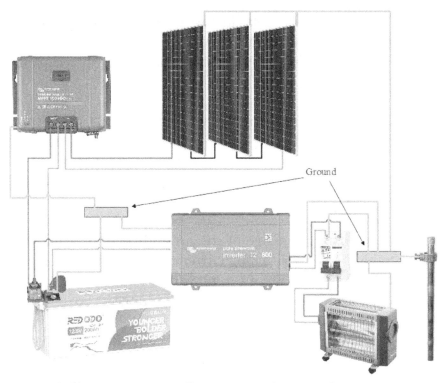

A diagram on grounding your stationary solar system

When grounding the frame of your solar panels, it's important not to place an additional grounding rod right next to them. If your house already has a grounding rod, putting another one nearby can create ground loops, leading to problems, especially if lightning strikes nearby.

Instead, run a grounding conductor, which should be the same size as your PV cables—directly to your main AC grounding busbar, which then connects to the main grounding rod of your system.

As previously mentioned, for your DC components (like the charge controller and inverter), you'll need a separate grounding busbar to bond them together.

Each DC component gets its own grounding wire going directly to the grounding busbar. Don't daisy-chain these components because if you remove one, you'll interrupt the grounding for the others.

From this grounding DC busbar, you should run a single cable to the negative terminal of the main battery. Or to the negative busbar if you have one. This is a one-time connection and should only be made once — **don't make this link it in multiple places**.

Now, about sizing the size of the grounding wire — this wire needs to be able to carry a fault current. There's some debate on the size, but a good rule of thumb is to use a conductor that's half the size of the current-carrying conductor. So if you're using 4 gauge wire (25mm^2) as your battery cables, use a 6 gauge grounding conductor, which is 16mm^2. Check your device manual for their specific recommendations as well.

On the AC side of your system, you'll have an AC ground coming out of the inverter. Your AC distribution box should have a grounding busbar. This is where you collect all the AC grounds from your loads. It's also where you'll bond the neutral and ground together (called a ground-neutral bond), which I will discuss soon.

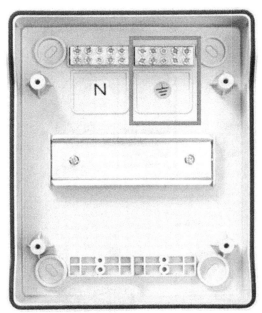

Grounding busbar in an AC distribution box (EU version)

Mobile Systems Grounding

Let's talk about mobile setups, like a camper or RV. In a mobile system, you don't have a grounding rod to stick in the ground. Instead, you'll connect your solar panel frame to the DC busbar, which is tied to your vehicle's chassis.

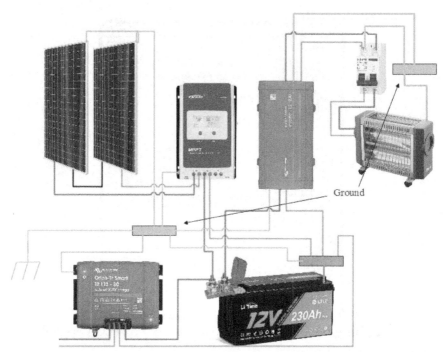

Ground

A diagram on grounding your mobile solar system

But before you do that, check the resistance between the roof racking (where your solar panels sit) and the chassis. If the resistance is zero, they're already connected through the metal racking, and you won't need to run an extra wire. Adding another wire could cause a ground loop, which we want to avoid.

As with the stationary system, all your DC components — like the charge controller, dc to dc charger, and inverter — should be bonded to the DC ground busbar. Then, run one cable from the busbar to the main battery negative terminal, and another to the chassis, which acts as the earth ground in mobile systems.

On the AC side, if you're using an inverter, you'll need to bond the neutral to the ground. But, if your inverter is also a charger, or an all-in-one system, this is a bit more complicated, and I will talk about it in the next chapter.

The AC devices will also have a grounding wire. Connect them to the AC grounding busbar located in the AC distribution box. This is similar to stationary systems.

Lightning

In the case of lightning, you need to connect all metal parts to the grounding busbar. If a metal pipe in a building is not grounded, and lightning strikes that pipe, it cannot find a way to go to earth. In this case, the electrons will try to find a way to another metal object. In this case, electrical arcing can happen, which can result in a fire. Or even worse, death to the person standing next to it.

This kind of grounding is called equipment grounding. Its purpose is to give electrons a path to the ground. Equipment grounding is not to protect your system from direct lightning, but it can reduce the chance of it. This is because the accumulated charges on your solar panels have a way to ground before lightning happens. Because the electrons are dissipated to the ground, there is less chance of lightning striking your panels because there is no charge on them.

| Signal | Chassis | Earth |
| ground | ground | ground |

Types of grounding (Wikipedia)
A grounding rod should be 8 feet long or 2.4 meters and have a diameter of 5/8 inches or 16mm². If you have dry soil, you need to add two or three rods, spaced 6 feet or 3 meters apart.

The grounding rod(s) resistance should be a maximum of 25 ohms. You will need a special tool called a 'clamp ground earth resistor tester'. These are expensive and should be measured by a professional if you want your system to pass inspection.

Whenever a lightning strike hits your solar panels, expect damage. Lighting protection is another subject and is not the same as equipment grounding.

You can protect your system with surge protectors or SPD's or lightning surge protection (LSP) from midnite solar, delta surge arrestors or littlefuse. Expect to pay around $130 each. There are cheaper ones online costing $15, but they don't have any certificates so I would not trust them.

I recommend using the midnite solar SPDs. Here is where to use them:

- **MNSPD-115:** Up to 90VAC circuits, 12V, 24V, 48VDC battery circuits.
- **MNSPD-300-AC:** 120/240 VAC circuits.
- **MNSPD-300-DC:** Off-grid PV combiners, charge controller inputs up to 300VDC.
- **MNSPD-600:** 316V/480 VAC circuits, grid-tie PV combiners, grid-tie inverter input, non-isolated inverters.

Neutral-ground bond

The neutral-ground bond is essential for two reasons:

- It ensures a stable voltage reference

- It allows ground fault current interrupters and circuit breakers to function correctly, protecting against ground faults, overcurrents, and electric shocks.

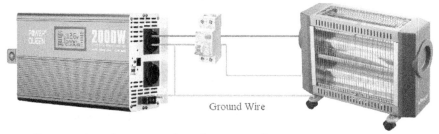

Ground Wire

Connecting the ground to the neutral on an inverter output

A neutral-ground connection (also known as a neutral-to-earth bond or MEN link) connects the neutral wire to the ground at a single point, typically at the main service panel, inside an inverter, or, in this diagram, at the inverter's AC out terminals.

This bond ensures that the neutral remains at 0 volts relative to the ground. Without this connection, the neutral will "float," leading to unpredictable voltages and increased electrical shock or equipment failure risks.

The neutral-ground connection also ensures that safety devices like ground fault circuit interrupters and circuit breakers work properly, as they rely on this bond to detect ground faults and prevent dangerous situations like electric shock.

How do you know if your inverter, inverter/charger, or hybrid inverter has a ground-neutral bond?

Let's see what you will measure without the bonding wire. If you are using inverters that have 230VAC output, you need to multiply these values by two.

- **Live to neutral:** 115V
- **Live to ground:** around 47V
- **Neutral to ground:** around 47V

These are the values we get when we make the neutral-to-ground connection and measure again.

- **Live to neutral:** 115V
- **Live to ground:** 115V
- **Neutral to ground:** 0V

You can see that the neutral to ground is now zero volts. This makes sense because they are bonded together. These are the values you will get when you measure your standard wall outlet (you can try it with your multimeter).

Some inverters make this bond internally. To check for a bond, you can measure the resistance between the neutral and ground connections. If the resistance is 0 or close to 0, the bond is already present. If it's high or shows an "open loop," there's no bond. I suggest turning off the inverter when you measure resistance.

Off-grid Systems

Lets explore two different systems. The first is a normal off grid system where there is only one ground to neutral bond. And the second is where an extra ground can be added in mobile systems through shore power.

In the following diagram, I manually created the bond with a jumper wire connecting the neutral and ground. I took the ground lug and connected a wire to the negative input of the GFCI (more on GFCI's later).

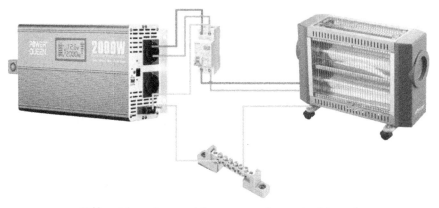

Off-grid system with a ground-neutral bond

Let's talk about a normal off grid system. In off-grid systems, the neutral-ground bond will be a permanent connection, because this will be the only place where a ground neutral connection can be made.

It can be located in the inverter or the distribution panel. This is the easiest system.

Measure if there is a ground-neutral bond and if not, create one. Connect the ground of your AC appliances to the AC grounding busbar and the grounding of the inverter. Then, connect one grounding wire from the inverter to the negative wire. In the diagram, I chose to do it at the input of the ground fault current interrupter, but you can connect it to the AC-out terminal block as well. Just do it before your ground fault current interrupter and breakers. Otherwise, they might not function properly.

So what should you do if you have multiple inverters in parallel? Only one inverter should create the neutral-ground bond to avoid grounding loops, otherwise it will cause issues.

Mobile systems

Now, mobile systems are a bit more complicated.
We are talking about a mobile system for a boat or an RV. It is more complicated because we can plug our system into shore power. This creates another connection to ground through the shore power connection.

Inverter only

Let's say you arrive with your RV at a campground and want to use your high-power appliances.

If you only use an inverter (no inverter/charger or hybrid inverter), you must use a transfer switch to choose between shore power or your own inverter. This is because your off-grid inverter cannot work with the grid.

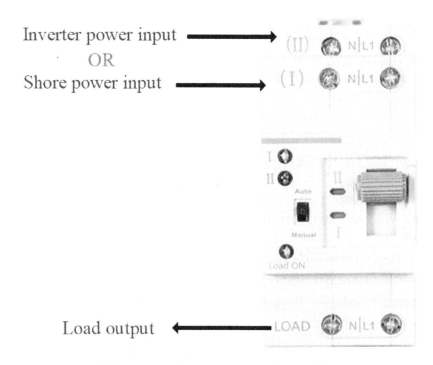

Inverter power input
OR
Shore power input

Load output

The connections on a transfer switch

The transfer switch let's you select which power source you want to use. So you will connect the shore power from the campground to the shore power input. And your inverter's output will be wired to inverter power input.

The load output will be used to power your loads while on the campground while your battery is being charged by the solar panels on your roof. During the time you are on the campground, you will use the electricity from the campground to run your appliances.

You can add a battery charger to recharge your household batteries so you have charged batteries when you leave the campground. The battery charger will be connected to the shore power input.

How does this work with the ground-neutral bond? Since the transfer switch doesn't have any grounding cable coming in, no additional grounding is brought into the system. So, this essentially works like an off-grid system, which we have discussed before.

Inverter/charger

If you are using an inverter/charger or an all-in-one inverter, you will have two different operational modes:

The first one is **inverter mode**, also called off-grid mode, for when you are driving on the road. The neutral-ground bond is created inside the inverter, similar to the off-grid systems we just talked about.

The second one is called **shore power mode**, which is connected to external power. This can be when you are on a campground plugged into shore power. When connected to shore power, you will have two neutral ground bonds in the system, which is undesirable.

The bond inside the inverter/charger must be disconnected because the shore power source already has a neutral to ground bond, and this will create grounding loops.

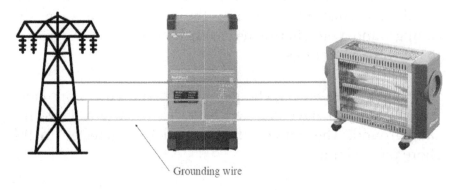

Grounding wire

As you can see, we now have two ground neutral bonds at the same time which we do not want. One inside of the inverter/charger, and one at the power utility. We have to disconnect the ground neutral bond inside of the inverter charger.

So, how do we remove the bond when we are connected to shore power? There are three different ways.

First option
The first one is manual bonding. You can manually create or disconnect the bond with a switch, but forgetting to do this could cause the ground fault current interrupter to fail, so I do not recommend this option.

Second option
The second one is when you don't have an inverter charger or an all-in-one inverter. If you are using a battery charger to charge the batteries, it doesn't add any additional grounding to the system.

However, this setup does not have AC pass-through, meaning you'll still run appliances on battery power, which may be insufficient for high-power devices like air conditioning units. To solve this, refer to the transfer switch option mentioned previously.

Third option
The last option is to use an inverter/charger or all-in-one inverter with a built-in grounding relay: the grounding relay automatically handles the bond depending on whether the system is in inverter or shore power mode. This is the preferred method.

The following schematic shows the simplified workings of a ground relay. When the grid is connected, contact 1 will be closed, and 2 will be open. This action enables the ground from the grid to be used while the ground from the battery is disabled.

When the grid is disconnected, contact 1 will be open, and contact 2 will be closed. Now, the ground will be the negative battery terminal.

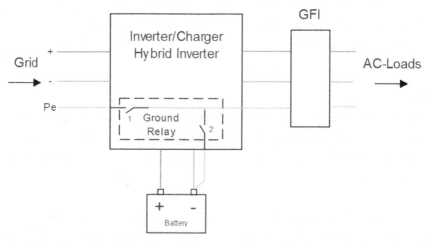

Simplified working of a grounding relay

As you can see in this image from Victron, when shore power is connected on the left, the AC input relay will close, forcing the earth relay to open automatically.

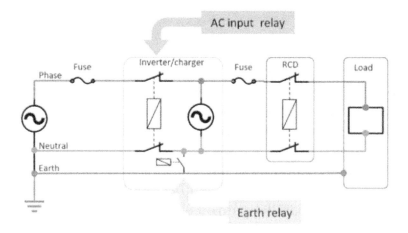

When the AC input is disconnected, the AC input relay opens, and the earth relay closes, creating the neutral ground bond. This is a fully automated process.

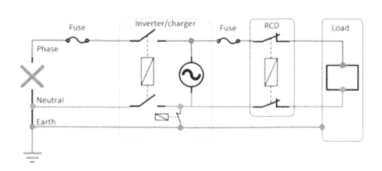

I suggest using an inverter/charger like the Victron MultiPlus for RVs or boats. It has a grounding relay, so you don't have to manage the bond or worry about it.

Again, if this sounds difficult, I recommend watching the video about it in this playlist:
https://cleversolarpower.com/book/3videos

GFCI

To be electrocuted, the current has to go through your body. There are different systems on how to protect yourself from being electrocuted. One misconception is that a circuit breaker will protect you from electrocution. This is not the case.

Another misunderstanding is that electricity takes the path of least resistance. This is not true either. Electricity will flow wherever it can in parallel paths. Let me explain this with an example.

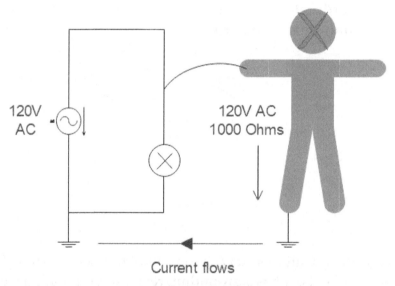

Current flows

Current flowing through a person

In this image, a person touches a live wire. The AC source is 120 volts alternating current. We know that in a parallel circuit, the volts are the same. The light will receive 120 volts, and the person touching the live wire will also receive 120 volts. In dry conditions, a person has a resistance of 1000 ohms from hand to feet. We apply Ohm's law to figure out how much current will pass through this person's body.

$$I = \frac{U}{R}$$

$$I = \frac{120V}{1000\Omega} = 0.12\,amps\ or\ 120mA$$

This will not end well for the person touching the live wire. The current will travel through the body and return to the source. Here are several current ratings and their consequences for the human body:

- 0.3mA: Sensation of touching a live wire.
- 0.7mA: Your let go threshold.
- 10mA: Max amount of current you can still let go of.
- 50mA: Fibrillation of the heart, which is fatal.

We want to protect ourselves from these deadly situations, so we are going to incorporate a GFCI (ground fault current interrupter). But first, we will discuss the different energy systems to understand why we should use a GFCI.

IT and TT Systems

There are several types of electrical systems. This book describes two networks that are the most common for off-grid use (the others are not discussed).

- IT: unearthed systems (Isolé-Terre or isolated earth)
- TT: earthed systems (Terre-Terre or earth to earth)

IT systems: The earth is not connected at the source. There is no connection between the negative and the grounding wire, so energy cannot flow back to the source.

The following diagram shows that the person touching one phase is safe from electrocution. This is because the path to the source is not completed (there is no grounding at the source). It is the same as birds sitting on a high-voltage exposed electric wire.

Now, if the person touches two phases, the circuit is completed. Current will flow through the person's heart, and the consequence will be fatal.

Consequences of touching a wire in IT system

This system is easy to install in off-grid applications. However, this is not the recommended system.

TT systems: The earth is connected to the negative wire, which in a mobile situation is the negative terminal of a battery.

If a person touches one of the live wires, the connection to the source will be completed. The current travels through the person to ground and back to the source, as shown in the following diagram. The person can die because the circuit is completed through the body. This is not what we want because it is more dangerous than the IT system (because the current can flow back to earth).

Therefore, we include a GFCI (ground fault current interrupter). This device detects any current leakage and will break the circuit as soon as it detects it. The device compares the current going into the circuit (+) and returning (-). These two must be in balance all the time. If the current goes back through the earth (if a fault occurs), the GFCI will detect this and break the circuit. This happens very fast in less than one-tenth of a second. It needs to act quickly so your body doesn't get electrocuted.

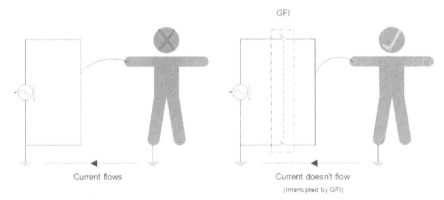

GFI

Current flows   Current doesn't flow
(Interrupted by GFI)

Consequences of touching a wire in a TT system

A GFCI has many different names. It is also known as a residual current device (RCD), earth leakage device, ground fault circuit interrupter, and residual current circuit breaker. All these names refer to the same device.

You might have noticed that the GFCI unintentionally breaks the circuit in your home. This is because there is a current leakage to the earth. It can be a fridge where the compressor is not working as intended, and the whole house is shut down.

It could also be a pump in your backyard where the GFCI trips because the pump has a small current leak to the earth. This is all very good in household applications because you know something is wrong.

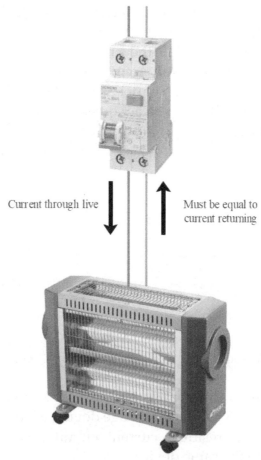

Current through live ↓   Must be equal to current returning ↑

Workings of a GFCI

However, this won't work in an IT system. Therefore, IT systems are used in industrial applications where one fault may not interrupt the current flow to keep machines working. The TT system is recommended for residential or off-grid purposes.

A GFCI works with AC, so it needs to be placed right after your inverter to protect you on the AC side of the system. You can only have one neutral-to-earth connection.

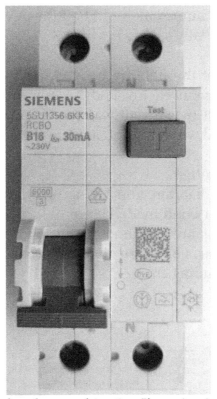

The GFCI + breaker combination I'm using in my videos (30mA from Siemens, DIN mounted)

In American homes, GFCIs are usually only installed in wet areas like bathrooms or outdoor outlets, where the risk of electric shock is higher (because the resistance is lower when your skin is wet).

However, in Europe, the GFCI protects the entire house. It's installed where the electricity enters the home, so every outlet and circuit in the house is protected from ground faults, not just in certain rooms. This ensures the whole house is covered.

One time, a GFCI saved my life. I was standing with my feet in the water in a pool filter room (I know, not very smart) while plugging something into a socket. The socket had a bad connection, and I was shocked. Thankfully, the GFCI detected the problem and immediately shut off the power, stopping the shock and potentially saving my life.

The current went through my body, straight into the ground (because water reduced my body's resistance to earth). The GFCI sensed that the current was not returning to its neutral phase but instead went through me into the ground.

Putting it all together.
In the following diagram, you can see an off-grid inverter with a ground neutral bond made before the ground fault current device. The appliance's ground conductor is wired to the AC grounding busbar. For instructions on grounding the system, refer to the grounding chapter.

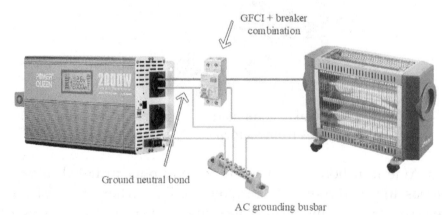

GFCI + breaker combination

Ground neutral bond

AC grounding busbar

Diagram of how to wire ground, make a neutral ground connection, and add a GFCI

Galvanic Corrosion

Imagine you have two different metals, like copper and steel, touching each other in water. Over time, you might notice that one of the metals starts to wear away or corrode. This is called galvanic corrosion. It happens because when different metals are in contact in a watery environment, they can start a tiny electrical current that causes one of the metals to corrode.

Why Does it Matter in Boats?

Boats often use various metals in their construction, always in contact with water. This makes them prime candidates for galvanic corrosion. If not managed, it can damage important parts of the boat, like the hull or electrical systems.

When a boat is connected to shore power, the earth is also connected. This can lead to a potential difference of several volts, which can cause galvanic corrosion.

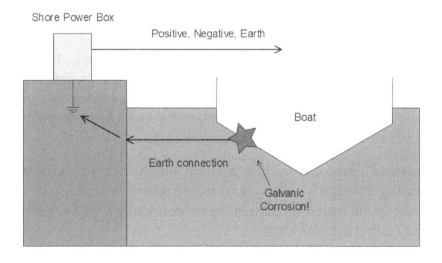

How to Prevent Galvanic Corrosion

1. **Galvanic Isolator**: This is like a gatekeeper for your boat's electrical system. It stops harmful electrical currents from the water from getting into your boat's system, which can cause corrosion. Think of it as a filter that only lets the good electricity through.

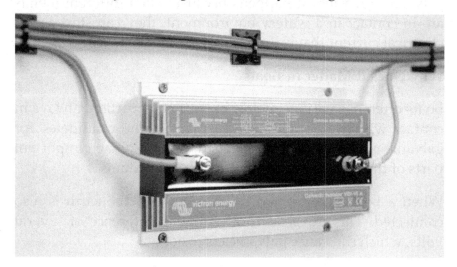

Galvanic isolator by Victron

2. **Isolation Transformer**: This is another tool for protection. It's like having a personal power station that separates your boat's electrical system from the shore power. This separation means the electrical currents that cause corrosion can't pass through.

Galvanic corrosion is like a silent enemy for boats, slowly causing damage without you noticing. Using tools like galvanic isolators or isolation transformers, you can protect your boat and keep it safe for longer.

Adding a Generator

Living off the grid is all about being self-sufficient, but sometimes nature doesn't cooperate as much as we'd like. That's where a generator comes in handy. Think of it like a backup plan. When your solar panels aren't getting enough sun to generate electricity, the generator saves the day. It's like having a spare tire in the trunk of your car - you might not need it all the time, but it's a lifesaver when you do.

Generators are also great for keeping your batteries charged. Imagine your batteries are like a water tank. If it hasn't rained for a while, the tank gets low. In the same way, if there's not enough sun, your batteries can run down.

Sometimes, you might want to use something that needs more power than your regular off-grid setup can handle. This could be a big power tool or several appliances at once. Your generator can provide that extra power boost, ensuring you can use these devices without a hitch.

Reliability is one of the best things about having a generator when living off the grid. Being off the grid means you need a dependable power source, and no matter what the weather's doing, you can count on your generator. Plus, if your solar or wind setup needs repairs and isn't making electricity, the generator can take over, so you're not left without power.

Starting with a generator can also be easier on your wallet. While setting up an extensive solar array and battery can be pretty expensive, a generator can be a more affordable way to get your off-grid life up and running. And in emergencies or unexpected situations, having a generator means you're prepared. It's like having a plan B for your electricity needs.

So, a generator is a trusty companion for anyone living off the grid. It ensures you have a steady and reliable power supply, ready to step in whenever your primary power sources aren't enough. It's all about giving you the power you need whenever you need it, no matter what.

Different Generator Systems

When adding a generator to an off-grid solar system for battery charging, you have two options: an integrated charger (usually part of an inverter/charger unit) or a standalone battery charger. Each option has unique considerations and benefits.

Before choosing your charging setup, it's essential to understand the difference between inverter generators and conventional generators:

- **Inverter Generators**: These are ideal for off-grid solar systems because they produce clean, stable power with minimal fluctuations, better for sensitive electronics, and efficient battery charging. They're also quieter and more fuel-efficient, especially at lower loads, as they can adjust their engine speed based on the power demand.

- **Conventional Generators**: These typically run at a constant speed regardless of load, which can lead to higher fuel consumption and noise. While effective, they're best suited for heavy, non-sensitive loads and may not be as efficient or stable when powering electronics.

When integrating a generator into an off-grid solar system, you have several configuration options based on your power needs, budget, and desired level of automation.

Inverter generator to a standalone charger

The generator directly powers the charger to charge the battery bank. This setup is relatively simple and cost-effective.

Inverter generators are well-suited for this configuration as they provide clean, stable power and adjust engine speed to match load demands, reducing fuel consumption and noise. This setup works well for users looking for an affordable and straightforward charging solution, especially if they don't require advanced programming features.

It is recommended that the generator be run at around 70% of its maximum load to optimize fuel efficiency and minimize noise. However, this setup requires manual starting of the generator, as it lacks the automatic features of an inverter/charger.

Example of a generator with a stand-alone charger

Inverter generator to an inverter/charger

In this configuration, the inverter generator supplies power to the inverter/charger, which manages battery charging and power distribution to AC loads.

This setup is ideal for users who want automated charging and the flexibility of AC power distribution directly from the inverter/charger. For example, your air conditioning can use 20A from the generator and 5A from your battery bank.

The inverter generator provides stable power, which is safe for sensitive electronics. The inverter/charger handles load balancing and automatically switches between solar, battery, and generator power, saving fuel and optimizing the generator workload. You also have the integrated ground-neutral bond, which is automated once the inverter/charger's AC input is activated.

Running the generator at about 70% of its maximum load is recommended for fuel efficiency and reduced noise. While this configuration provides a higher level of automation and supports complex off-grid systems, it generally requires some programming to set charging parameters and remote start options if available. Victron recommends these generator sizes for their inverter/chargers:

Inverter Charger Size	Minimum Recommended Generator Size	240V AC Input Current Limit
800VA	2 kVa	5A
1200VA	2.5 kVa	7A
1600VA	3 kVa	10A
3000VA	5 kVa	16A
5000VA	8 kVa	25A

Recommended sizing by Victron (note: in VA)

For a complete FAQ about generators and Victron equipment, visit the following URL:
https://www.victronenergy.com/live/multiplus_faq

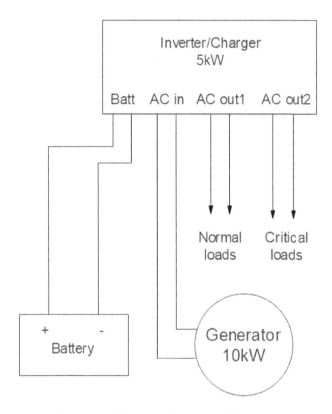

Diagram of inverter/charger with generator

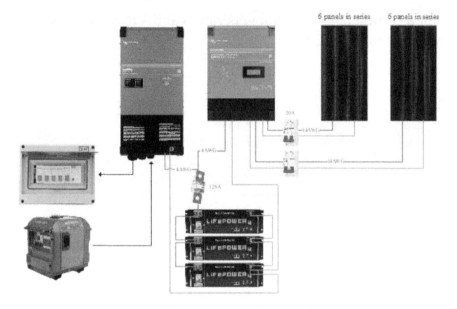

Example of a diagram with generator and inverter/charger from Victron

Conventional generator with a standalone charger

This configuration uses a conventional generator to power a standalone charger, which connects directly to the battery bank. It's an economical choice and suits users who need a reliable but basic solution for occasional use.

Conventional generators typically lack load-responsive speed adjustment, leading to higher fuel consumption and more noise than inverter generators, especially when running below full load.
This setup provides flexibility in choosing a charger matched to the battery bank's voltage, but conventional generators offer less stable power, making them less ideal for sensitive electronics or efficient battery charging.

Running a conventional generator at about 75% load is recommended to avoid overloading, although they won't adjust engine speed as an inverter generator would. Additionally, users will need to monitor charging manually and switch between power sources, as this setup doesn't offer the automation provided by an inverter/charger.

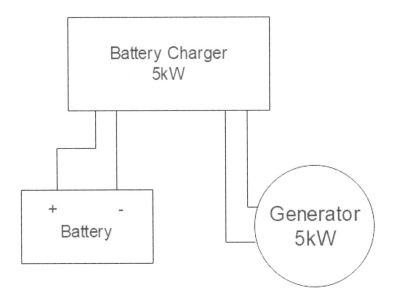

Diagram of generator with charger

You need an additional charger for this. The AC output from the inverter needs to be adjusted to the battery's voltage. Some good chargers are from EG4. They have several chargers for different voltages. A good example is their 48V 100A chargeverter.

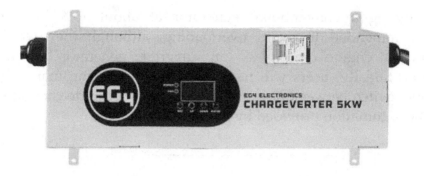

EG4 48V 100A chargeverter

You can program the charging current in this device. To reduce noise and save fuel, it should be a maximum of 75% of the generator's continuous rating.

If your generator is 5kW, you should charge the battery with 3.75kW (5kW*0.75=3.75kW). The current you should program will be 3750W/56V=67A

To enable automatic starting, a generator with a remote start function with the inverter/charger is optional. These are usually more expensive, like the Kohler brand.

Adding a generator to an inverter/charger will require some programming work in the device. If you are less tech-savvy, you might consider the first option.

Calculating charging time

We have a 5kWh battery (48V 100Ah server rack) and a 2kW generator. We will charge the battery using a standalone charger.

We must consider a few factors to calculate how long it would take to recharge a 5 kWh battery using a 2 kW (kilowatt) generator. The most important are:

- The capacity of the battery
- The output of the generator
- The recommended load on the generator
- The efficiency of the charging process

Combining all these parameters gives us the following formula:

$$\frac{Battery\ capacity\ in\ Kwh}{Generator\ power \times Recommended\ power \times efficiency}$$

$$\frac{5kWh}{2kW \times 0.75 \times 0.8} = 4.1\ hours$$

It would take approximately 4.17 hours, or about 4 hours and 10 minutes, to fully recharge a 5 kWh battery using a 2 kW generator, considering a 75% recommended load and 80% charging efficiency.

DIY Solar Power Setup

The procedure for building a solar power system is straightforward and can be done by following these steps:

1. Calculations and schematic

Before you order components or even consider making a solar system, you need to do the calculations first. Refer to the chapter 'Sizing your Solar System' for more information.

Drawing out your solar system will make it easy to assemble the components later. It will also give you an estimation of the space and components you will need.

It will look overwhelming in the beginning. What helps is seeing the system as modules. First, you calculate the number of solar panels, calculate the wires and voltage drop, and decide on the charge controller. Then, you move on to the inverter cable sizing, etc. Seeing components as modules that will be added together will make it easier for you to draw the system.

You can use the diagrams on my website as inspiration. Check them out here: https://cleversolarpower.com

2. Ordering and preparing components

After you have done the load analysis and drawn a schematic, it's time to order your components. To avoid unnecessary purchases, make sure you have read all the chapters in this book before buying your components.

Remove the components from their boxes and place them roughly where you want them. Mounting the components on a fire-resistant board is essential. Make sure to put them as close together as possible to reduce wire losses and save on wire costs (especially with thick wire). Leave some space for airflow. Make sure the connections are accessible for future upgrades or maintenance.

3. Layout

Look for a space to place your batteries. Remember to respect the limits and parameters stated in the National Electrical Code section of the book. If you have the option, make sure the batteries are placed at room temperature.

Place the inverter and the charge controller in an ideal position close to the battery bank. This reduces the voltage drop and allows for easier installation.

4. Wiring the batteries

Before wiring your batteries, remember to review the section related to series and parallel connections in the battery section of this book.

Choose your wire gauge wisely to handle the current flow from your charge controller to the batteries. Check the charge controller's manual for the recommended wire thickness.

Add a fuse as close to the positive terminal as possible (I recommend an MRBF). Add a battery bank disconnect switch (high-current, low-DC voltage switch).

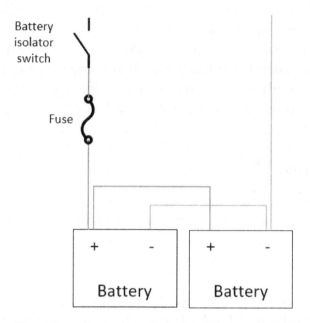

Two batteries in parallel with fuse and switch

If you use a shunt, now is the time to wire it in. The shunt's negative terminal will act as the main negative battery terminal.

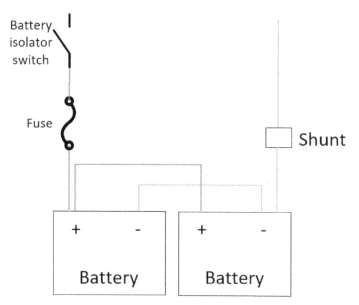

Shunt at the negative battery terminal

5. Wire the battery to the charge controller.

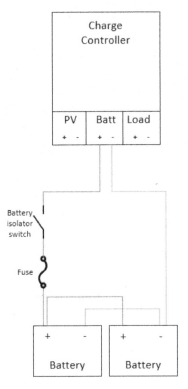

Wiring the batteries to the charge controller

Once wired, the charge controller's display will light up. If you have a Bluetooth charge controller, now is the time to configure it in the app. Select the battery type you are using and follow the directions in the manufacturer's manual to do this.

6. Install the inverter

You have two choices when installing the inverter:

- Wiring directly to the battery terminals.
- Wiring from a busbar.

Wiring from the battery terminals:

This option is easier than wiring from the busbar because you need fewer cables. The downside is that it is harder to expand this setup later. Remember that the maximum number of connectors on a terminal is three.

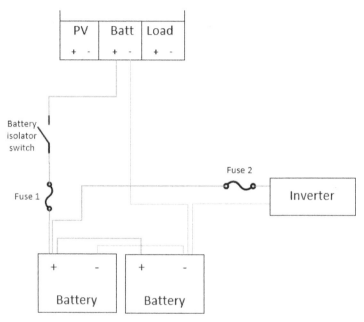

Wiring the inverter to the batteries

Fuse 2 needs to be able to handle the current that the inverter draws. Select the wire diameter and choose the correct fuse for it. You should install a busbar if you do not want to wire directly from the batteries.

The following image shows how to wire from the busbar. It shows that the charge controller is wired straight to the battery, but it can also go to the busbar.

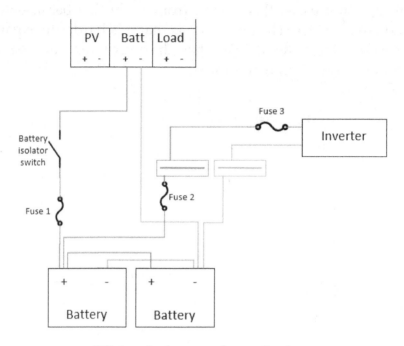

Wiring the inverter from a busbar

In this case, the size of fuses 2 and 3 will be the same. If no other loads are attached, it's enough to use only fuse 2. However, if you expand your system with a DC fuse box, the size of fuses 2 and 3 will not be the same.

If you choose an inverter/charger, install the AC input to accept shore power or power from a generator or the grid. You can install a two-way switch (called a changeover switch) or transfer switch to choose between generator or shore charging.

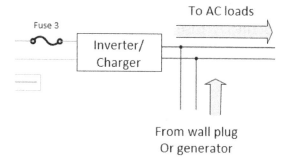

To AC loads

Fuse 3

Inverter/
Charger

From wall plug
Or generator

Installation of inverter/charger

Selection between shore and generator
(image from Victron multiplus diagrams)

7. DC fuse box

You can connect the DC fuse box using the load terminals of the charge controller.

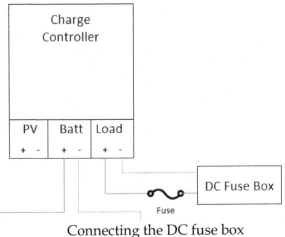

Connecting the DC fuse box

However, if the output terminals can only supply 20 Amps (depending on the model), you will be limited to 12 *Volts* × 20 *Amps* = 240 *Watts*.

This can be a problem, especially if you have many DC loads. Therefore, it is recommended that you connect the DC fuse box to the previously installed busbar or the battery terminals.

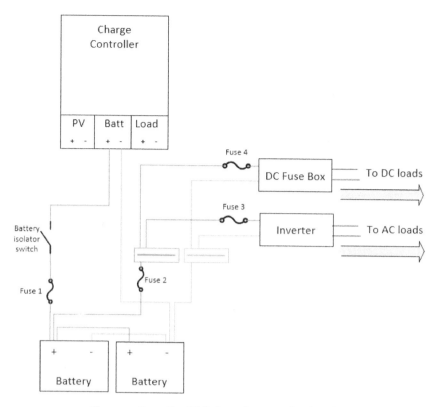

Connecting the DC fuse box to the busbar

In this case, fuses 2 and 3 won't be the same. Using your load estimation, you must calculate the maximum AC and DC loads and select the correct fuse for each. Add up fuses 3 and 4 to know the value of fuse 2. Select your wire according to the maximum amount of current that goes through the wire.

8. Install the solar panels

Refer to the chapter about mounting your solar panels. Plan your solar panels so the wire to the charge controller is as short as possible to limit voltage drop. Connect your solar panels with steel wire if you put them on a vehicle's roof. If one comes loose, it won't fly off.

9. Wire the modules

Read the chapter 'Series and Parallel Connection' to understand how to wire the modules for your intended purposes.

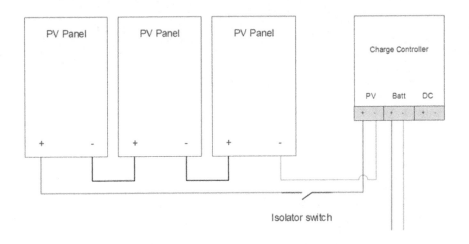

Connecting three panels in series

If maintenance is required, the solar string can be disconnected using the solar isolator switch. Try to place the switch in a location where you can easily reach it, preferably close to the charge controller. You can also use a DC breaker mounted on a DIN rail. This will be cheaper than a solar disconnect switch.

The cable entry plate allows you to safely connect the output of the solar panels from the outside to the inside of a wall or vehicle. Use enough sealant to protect your wall or roof from letting in water during rain.

10. Install the B2B charger

If you would like to charge the solar battery using the car alternator, install the battery-to-battery charger. Refer to the wiring scheme in the 'B2B charger' chapter. It's best to place your house batteries as close to the alternator as possible (behind the driver's seat). This will reduce the wire length. Otherwise, you will need thicker cables for the voltage drop because the B2B charger will have a low voltage output, a high current, and a long wire.

11. Testing

Congratulations, you have completed the solar system installation. Now, you must run some tests. Testing includes:

- Checking for loose wires.
- Checking for sharp edges that can cut your cables.
- Monitoring the temperature of components.
- Monitoring the temperature of the wires.
- Checking the battery voltage when fully charged.
- Testing loads.
- Checking your wire connections at least once a year.

Solar System Examples

This chapter presents some examples and explains the reasoning behind wire thickness and fuses. It will be helpful to refer to this information when making your system.

Here are the systems we will discuss:

- 12V 500W inverter with 400W of solar

- 24V 1kW inverter with 800W of solar

- 48V 3kW inverter with 3kW of solar

- 48V 5kW inverter with 9kW of solar

Note about cable sizing: The cables in these systems are copper and are calculated with an insulation temperature of 75°C or 167°F. Do not use cables lower than this temperature if you plan to copy these systems.

The blueprints of these systems will be available on the website, https://cleversolarpower.com/offgridsolarbook, where you can view them in color and in larger sizes.

12V 500W inverter with 400W of solar

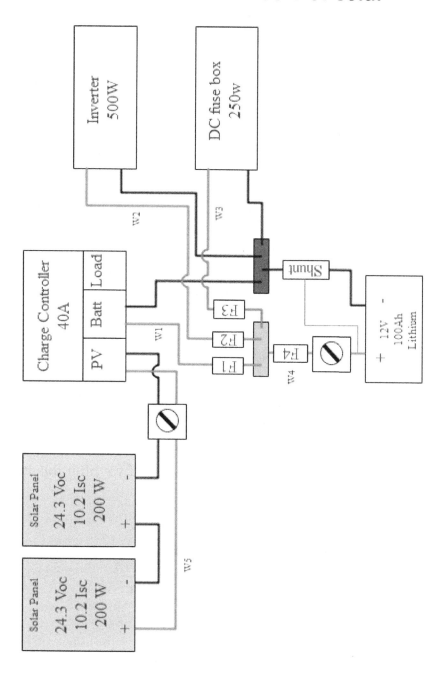

This system is suited for small loads no larger than 500W AC and 250W DC. Both loads can be active at the same time. Two solar panels, totaling 400W, charge the battery. Both 200W panels will be wired in series to the charge controller (Epever Tracer 4210AN), which charges the battery at 33A.

$$\frac{400W}{12V} = 33.33A$$

Next, I'm going to explain the wire sizes and fuses.

F1: We don't need a fuse here. The function we need here is a switch to disconnect the solar panels from the system. A solar disconnect switch or a DC breaker can accomplish this. Let's calculate the current value of the DC breaker. The Isc (current short circuit) of this 200W panel is 10.2A, and the Voc (volt open circuit) is 24.3V. The maximum current through this wire will be:

$$10.2A \times 1.56 = 16A$$

Therefore, we will use a 20A breaker, which is the size Rich Solar (the solar panel manufacturer) recommends.

Next, we will calculate the wire itself. We assume the charge controller's wire length is 20ft or 6 meters. If we plug this in a voltage drop calculator:
https://cleversolarpower.com/book/voltagedrop

Let's say the Vmp is 18VDC, we become 12AWG or 4mm² with a voltage drop of 2%.

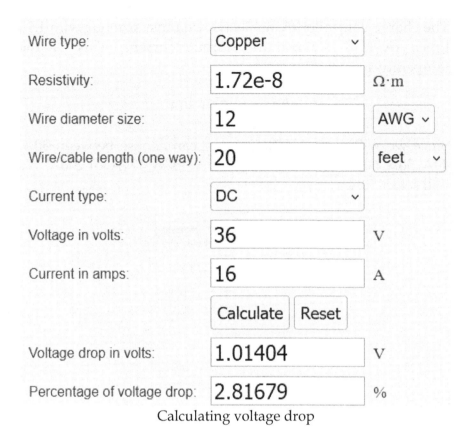

Wire type:	Copper ⌄	
Resistivity:	1.72e-8	$\Omega \cdot m$
Wire diameter size:	12	AWG ⌄
Wire/cable length (one way):	20	feet ⌄
Current type:	DC ⌄	
Voltage in volts:	36	V
Current in amps:	16	A
	Calculate Reset	
Voltage drop in volts:	1.01404	V
Percentage of voltage drop:	2.81679	%

Calculating voltage drop

F2: The following wire goes from the charge controller to the busbar. The maximum current that can go through this wire is:

$$\frac{400W}{12V} = 33.3A$$

A safety factor of 1.25 needs to be applied:

$$33.3A \times 1.25 = 41.6A$$

This is over the charge controller's max output. We will have a small loss when the conditions are better than STC. This will be very limited, so we will use the 40A charge controller. The controller will not be damaged, but it will be limited to 40A.

The charge controller manual says we must size our wire 1.25 times over the maximum output current. This is the temperature correction factor.

$$40A \times 1.25 = 50A$$

Since we will use a MIDI fuse, we can choose between 30A, 40A, 50A, 60A, 70A, 80A, 100A, 125A, 150A, 175A, or 200A. We will select 50A.

Current Rating (A)	Housing Material Color
30	▬
40	▬
50	▬
60	
70	▬
80	☐
100	▬
125	▬
150	▬
175	▬
200	▬

MIDI fuse available current ratings from littlefuse

I prefer using MIDI fuses for smaller systems because they offer a wider range of lower current ratings. MEGA fuses, on the other hand, start at 40A and then jump to 60A, 80A, 100A, and so on, providing fewer options in the lower current range.

A wire rated at 75°C or 167°F that can carry at least 50A is a 6AWG or 16mm².

SIZE AWG OR kcmil	Copper Conductors		
	Temperature Rating of Conductor		
	60°C	75°C	90°C
	TYPES TW UF	TYPES RHW THW THWN / THHW XHHW USE	TYPES RHH RHW-2 XHHW XHHW-2 XHH / THHW THWN-2 THW-2 THHN USE-2
14**	20	20	25
12**	25	25	30
10**	30	35	40
8	40	50	55
6	55	65	75
4	70	85*	95*

Selecting wire size

F3: This wire goes from the busbar to the inverter. The inverter has a power rating of 500W. That means we have a current of:

$$\frac{500W}{12V} = 41.6A$$

We must multiply this by a safety factor of 1.25, so the new current will be 52A.

$$41.6A \times 1.25 = 52A$$

The closest MIDI fuse to 52A is 60A. Now, we need to find a cable that can carry at least 60A because the fuse must blow first before the wire gets too hot.

Again, a wire that can carry 60A is 6AWG or 16mm².

SIZE AWG OR kcmil	Copper Conductors					
	Temperature Rating of Conductor					
	60°C	75°C		90°C		
	TYPES	TYPES		TYPES		
	TW UF	RHW THW THWN	THHW XHHW USE	RHH RHW-2 XHHW XHHW-2 XHH	THHW THWN-2 THW-2 THHN USE-2	
14**	20	20		25		
12**	25	25		30		
10**	30	35		40		
8	40	50		55		
6	55	65		75		
4	70	85*		95*		

Selecting wire size

F4: This wire goes from the busbar to the DC fuse box. Let's assume the fuse box will have a max load of 250W. That means this wire needs to carry the following current:

$$\frac{250W}{12V} = 20.8A$$

We multiply with a safety factor of 1.25 and become 26A. The closest fuse is 30A. A cable that can carry at least 30A is 10AWG or 6mm².

F5: This wire goes from the busbar to the battery. The maximum current this wire will carry is the sum of the inverter and the DC fuse box. This is

$$\frac{750W}{12V} = 62.5A$$

After the safety factor of 1.25, we become 78A. The fuse closest to this is 80A. A cable that can carry at least 80A is 4AWG or 25mm².

A battery isolator switch is between the busbar and the battery. This switch isolates the battery from the system if maintenance is required.

A shunt is on the negative connection to the battery. This is to monitor the state of charge of the battery. It is a much more accurate way of telling a lithium battery's state of charge (SOC).

I recommend using a fuse holder with a busbar for MIDI fuses; this will make installation a lot easier.

MIDI fuse busbar

24V 1kW inverter with 800W of solar

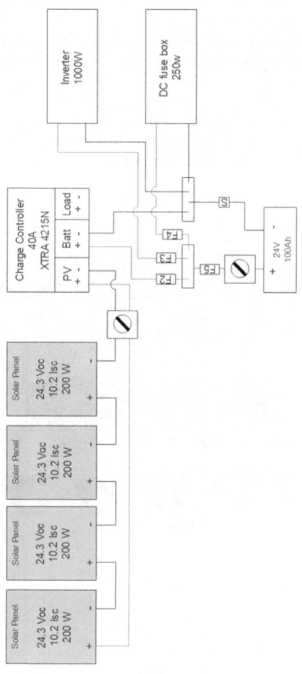

This system is suited for loads no larger than 1000W AC and 250W DC. Both loads can be active at the same time. The battery gets charged with a total of 800W of solar. Four 200W panels will be wired in series to the charge controller, which charges the battery at 33A. As you can see, we have doubled the solar power, but the charge controller stays the same because we have doubled the battery voltage.

The capacity (Ah) of the battery is 100Ah. This means that we can discharge a lithium battery at 1C at 100Amps. This will be enough for this system. If we have a 24V lead-acid battery, I do not advise you to use this inverter size as it will reduce the lifespan of your batteries. The 0.2C discharge rate of lead-acid only allows for:

$$100Ah \times 0.2C = 20A \text{ charge/discharge current}$$

$$20A \times 24V = 480W$$

As we will see in this example, the max discharge current is 52A. This system will work with lead acid, but it's not recommended.

F1: The wire goes from the solar panels to the charge controller. The Isc of this 200W panel is 10.2A, and the Voc is 24.3V (same as the previous example). The maximum current through this wire will be:

$$10.2A \times 1.56 = 16A$$

We will use a solar disconnect switch or a 20A breaker. This is the same as the previous example. The difference here is the voltage. The voltage will be 122V DC to the charge controller:

$$24.3Voc \times 4panels \times 1.25 = 121.5V$$

This is over the previous charge controller's maximum input voltage. We must use a charge controller with a maximum input voltage of 150VDC. We will use the Epever XTRA series 40A or the Victron 150/35.

Next, we will calculate the wire. We assume the charge controller's wire length is 20ft or 6 meters. If we plug this data in a voltage drop calculator with the Vmp being 18VDC per panel, we become 14AWG or 2.5mm² with an acceptable voltage drop of 2.2%.

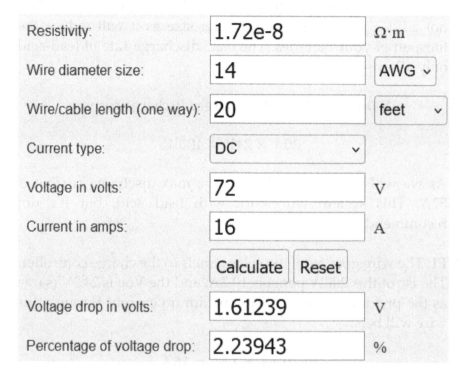

Resistivity:	1.72e-8	Ω·m
Wire diameter size:	14	AWG ⌄
Wire/cable length (one way):	20	feet ⌄
Current type:	DC ⌄	
Voltage in volts:	72	V
Current in amps:	16	A
	Calculate Reset	
Voltage drop in volts:	1.61239	V
Percentage of voltage drop:	2.23943	%

Link to the calculator:
https://cleversolarpower.com/book/voltagedrop

F2: The following wire goes from the charge controller to the busbar. The maximum current that can go through this wire is:

$$\frac{800W}{24V} = 33A$$

However, we should calculate it on the charge controller's maximum output current. The Epever charge controller has a limited output of 40A. The manufacturer says we need a safety factor of 1.25 times this current.

$$40A \times 1.25 = 50A.$$

Like the previous example, we will use MIDI fuses, and the closest MIDI fuse to 50A is 50A. A wire suitable to carry more than 50A is a 6AWG or 16mm². An 8AWG cable is not enough because it can only carry 50A, the same as the fuse. We need a wire that can carry more than 50A.

	Copper Conductors		
	Temperature Rating of Conductor		
SIZE	60°C	75°C	90°C
AWG OR kcmil	TYPES TW UF	TYPES RHW THW THWN	THHW XHHW USE	TYPES RHH RHW-2 XHHW XHHW-2 XHH	THHW THWN-2 THW-2 THHN USE-2
14**	20	20	25
12**	25	25	30
10**	30	35	40
8	40	50	55
6	55	65	75
4	70	85*	95*

Selecting wire size

F3: This wire goes from the DC busbar to the inverter. This wire needs to be kept short. Because we have a 24V system, we have a maximum current of:

$$\frac{1000W}{24V} = 41.67A$$

We need a safety margin of 1.25, so the current becomes 52A. The closest fuse to 52A (always round up) is a 60A MIDI fuse. A cable that can carry at least 60A is a 6AWG or 16mm²cable.

F4: This wire goes from the DC busbar to the DC fuse box. The maximum power that will be used is 250W. This gives us a current of:

$$\frac{250W}{24V} = 10.41A$$

Applying a safety factor of 1.25, we become 13A. the closest fuse to 13A is a 30A MIDI fuse, because this is the lowest value. A wire that can carry at least 30A is a 10AWG or 6mm² cable.

	Copper Conductors		
	Temperature Rating of Conductor		
SIZE	60°C	75°C	90°C
AWG OR kcmil	TYPES TW UF	TYPES RHW THHW THW XHHW THWN USE	TYPES RHH THHW RHW-2 THWN-2 XHHW THW-2 XHHW-2 THHN XHH USE-2
14**	20	20	25
12**	25	25	30
10**	30	35	40
8	40	50	55
6	55	65	75
4	70	85*	95*

F5: This wire goes from the battery to the DC busbar. It will be the wire with the most current in the system. If we add up all the power in the system (inverter + DC fuse box), we get 1250W. We divide the power by the battery's voltage to get the total current.

$$\frac{1250W}{24V} = 52A$$

With a safety factor of 1.25, we become 65A. A fuse close to 65A is a 70A MIDI fuse. A wire that can carry at least 70A is 4AWG or 25mm². We need to keep this wire as short as possible to minimize voltage drop.

SIZE AWG OR kcmil	Copper Conductors		
	Temperature Rating of Conductor		
	60°C	75°C	90°C
	TYPES	TYPES	TYPES
	TW UF	RHW THHW THW XHHW THWN USE	RHH THHW RHW-2 THWN-2 XHHW THW-2 XHHW-2 THHN XHH USE-2
14**	20	20	25
12**	25	25	30
10**	30	35	40
8	40	50	55
6	55	65	75
4	70	85*	95*

Finding wire size

Using a MIDI fuse busbar like the previous example is recommended.

48V 3kW inverter with 3kW of solar

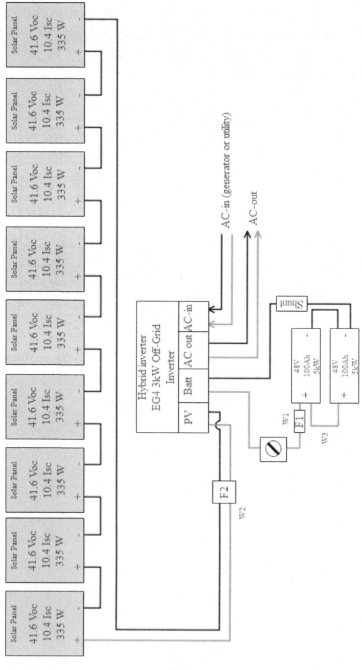

This system could be used for small to medium off-grid homesteads. We have chosen nine 335W solar panels from Rich Solar ($0.68/W). These panels have the following specifications:

- Maximum Power(Pmax): 335W
- Open Circuit Voltage(Voc): 41.6V
- Maximum power voltage (Vmp): 34.1V
- Short Circuit Current(Isc): 10.4A
- Max Series Fuse Rating: 20A

We connect these solar panels to the off-grid inverter from EG4. This is a 3kW hybrid inverter that includes an MPPT charge controller with a max input voltage of 500 VDC. We can connect all our solar panels in series to this hybrid inverter.

Connecting this hybrid inverter to the utility or a generator is possible.

Lastly, we have chosen two LiFePO4 server rack batteries. These can be from EG4, SOK, or others. We have selected two at 48V and 100Ah. They have a combined 10kWh of storage.

The 9 solar panels have a combined power output of 3,015W. If you live in a place with three sun hours per day, you can charge these batteries in one day.

$$3,015W \times 3hours = 9.045Wh \ or \ 9kWh$$

The wire size from the solar panels to the hybrid inverter will have a maximum current of 10.4A. We need to apply the safety factor of 1.56.

$$10.4A \times 1.56 = 16.2A$$

The solar panel voltage will be:

$$41.6 Voc \times 9\,panels \times 1.25 = 468 VDC$$

If the distance from the solar array to the inverter is 50 feet (15 meters), we could initially consider using 16 AWG (1.5 mm²) wire. However, 16 AWG isn't rated to carry 16.2A with 75°C insulation, so we need to increase the wire size to 14 AWG. Since PV cable doesn't come in 14 AWG, the closest available size is 12 AWG (4 mm²), which we'll need to use as the minimum size for this setup.

$$34.1V \times 9\,panels\,in\,series = 307V$$

Wire type:	Copper ⌄	
Resistivity:	1.72e-8	Ω·m
Wire diameter size:	16	AWG ⌄
Wire/cable length (one way):	50	feet ⌄
Current type:	DC ⌄	
Voltage in volts:	307	V
Current in amps:	16.2	A
	Calculate Reset	
Voltage drop in volts:	6.48963	V
Percentage of voltage drop:	2.11389	%

Voltage drop calculation

The DC breaker on this wire should be 20A because this is the fuse size recommended in the solar panel specifications. If you don't want the breaker, you can use a solar disconnect switch rated for 500VDC.

The following wire goes from the hybrid inverter to the server rack batteries. The batteries are wired in parallel. The main positive comes from battery one, and the main negative comes from battery two. This ensures the current is shared between the two batteries, and one doesn't work more than the other. A battery busbar will be a good option if you have more than two batteries.

The maximum amount of current through this wire is:

$$\frac{3,000W}{48V} = 62.5A \times 1.25 = 78.5A$$

We cannot use MEGA fuses since we use lithium batteries at 48V and 200Ah. We need to use a Class-T fuse or NH00 for the battery fuse.

A class-T fuse of 100A will keep this wire safe from overcurrent.

If the server rack battery has an integrated breaker, you don't need to add a fuse. Ensure the breaker is rated correctly according to the current you will draw and capable of breaking a short circuit current of 200Ah x 10 = 2,000A at 56VDC minimum.

The standard breaker current of a 48V 100Ah battery is 125A. Why? The battery can deliver 1C of current at 100Ah; this is 100A. If you add the safety factor of 1.25, this becomes 125A.

Since we have chosen a 100A fuse, we must select a cable that can carry at least 100A. Let's use a welding cable rated at 105°C or 221°F.

PowerFlex Welding & Battery Cable Specifications

Part # BL = Black RD = Red	Size (AWG)	Stranding (0.25mm)	Insulation Thickness (inches)	Conductor Diameter (inches)	Approx. Total Diameter (inches)	Maximum Amperage	Approximate Weight (lbs/ft)
CBL-CSBL-08 CBL-CSRD-08	8	131	0.07	0.129	0.31	55	0.07
CBL-CSBL-06 CBL-CSRD-06	6	263	0.07	0.20	0.36	115	0.11
CBL-CSBL-04 CBL-CSRD-04	4	370	0.07	0.23	0.40	150	0.15
CBL-CSBL-02 CBL-CSRD-02	2	634	0.07	0.30	0.46	205	0.24
CBL-CSBL-10 CBL-CSRD-10	1/0	1004	0.08	0.37	0.56	285	0.38
CBL-CSBL-20 CBL-CSRD-20	2/0	1255	0.08	0.43	0.63	325	0.45
CBL-CSBL-40 CBL-CSRD-40	4/0	2047	0.08	0.56	0.75	440	0.72

Selecting welding cable from Windynation

We can see that we need a 6AWG or 16mm² cable. To keep it simple, you can size the interconnecting cable to the other battery at the same size.

I recommend using a breaker at the AC-out/ It is calculated as follows.

$$\frac{3,000W}{120VAC} = 25A \times 1.25 = 31A$$

We should use a breaker at 32A. A cable that can carry at least 32A is 10AWG or 6mm². From there, you should wire it into a GFCI and then the breakers.

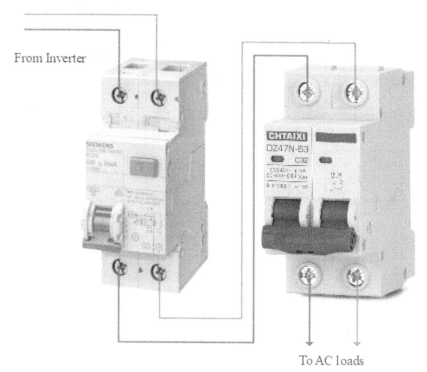

From Inverter

To AC loads

Wiring the AC side

The ground neutral bond is located inside the hybrid inverter, and you must manually make it. The manual provides instructions on how to do this.

48V 5kW inverter 9kW of solar

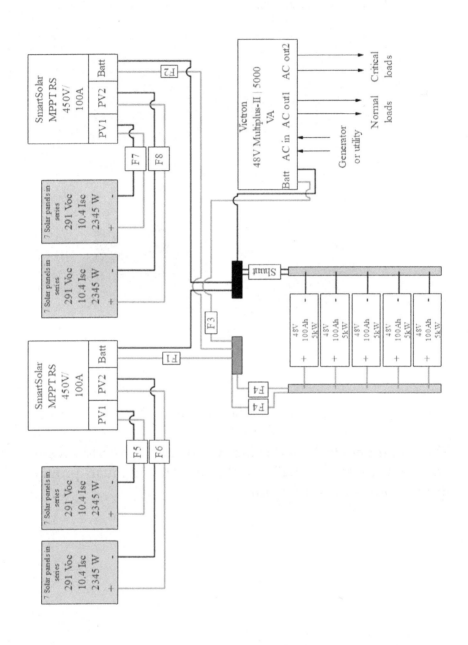

This system is for a large off-grid homestead or a house. We have chosen 28 solar panels from Rich Solar. These panels are the same as before and have the following specifications:

- Maximum Power(Pmax): 335W
- Open Circuit Voltage(Voc): 41.6V
- Maximum power voltage (Vmp): 34.1V
- Short Circuit Current(Isc): 10.4A
- Max Series Fuse Rating: 20A

This time, instead of using an all-in-one hybrid inverter, we will choose separate components. We will use two 450V/100A charge controllers and one Victron Multiplus II 5kW inverter. You can integrate this inverter with your home. When a blackout occurs, you still have power for your critical loads.

You can add more units in parallel if you need more inverter power. We will use 5x 48V 100Ah server rack batteries. These are 25kWh of storage, and if you have 3 sun hours, the solar panels will charge them in one day.

$$9,380W\ solar\ \times 3hours = 28Kwh$$

Now we need to calculate the voltage at the input of the charge controllers. This may not exceed 450V.

$$41.6Voc\ \times 7panels\ \times 1.25\ = 364V$$

The solar panels will be divided into 4 parallel arrays with 7 panels in series (7S4P). This will give us a total combined power of 9,380W of solar.

The current through this wire is:

$$10.4A\ \times 1.56\ = 16.2A$$

Like the previous system, as the solar panel manufacturer recommended, we need to use a breaker or solar disconnect switch (depicted on the diagram) on every series string to disconnect it from the system (F5,6,7, and 8).

Let's calculate the voltage on this cable:

$$34.1Vmp \times 7 = 238.7V$$

Let's say the wire is 100 feet or 30 meters long. Then we need a 12AWG or 4mm² wire.

Wire type:	Copper ⌄	
Resistivity:	1.72e-8	Ω·m
Wire diameter size:	12	AWG ⌄
Wire/cable length (one way):	100	feet ⌄
Current type:	DC ⌄	
Voltage in volts:	238	V
Current in amps:	16.2	A
	Calculate Reset	
Voltage drop in volts:	5.13359	V
Percentage of voltage drop:	2.15697	%

Link to calculator:
https://cleversolarpower.com/book/voltagedrop

The following wire will go from the charge controller to the busbar (F1+F2). The maximum current through the charge controller is 100A. We need to apply the safety factor:

$$100A \times 1.25 = 125A$$

Let's use MEGA fuses for this system. However, this is a 48V system, so we cannot use regular MEGA fuses, which are only rated to 32V. We have to use MEGA fuses rated at 70V. This is the table with the available sizes:

Part Number	Current Rating (A)	Color Code
0998060._	60	
0998080._	80	
0998100._	100	
0998125._	125	
0998150._	150	
0998175._	175	
0998200._	200	
0998225._	225	
0998250._	250	
0998300._	300 ¹	
0998350._	350 ¹	
0998400._	400 ¹	
0998450._	450 ¹	
0998500._	500 ¹	

70V rated MEGA fuses

We can use a 125A fuse. Let's use welding cables rated for 105°C or 212°F insulation temperature. A cable that can carry at least 125A is 4AWG or 25mm².

PowerFlex Welding & Battery Cable Specifications

Part # BL = Black RD = Red	Size (AWG)	Stranding (0.25mm)	Insulation Thickness (inches)	Conductor Diameter (inches)	Approx. Total Diameter (inches)	Maximum Amperage	Approximate Weight (lbs/ft)
CBL-CSBL-08 CBL-CSRD-08	8	131	0.07	0.129	0.31	55	0.07
CBL-CSBL-06 CBL-CSRD-06	6	263	0.07	0.20	0.36	115	0.11
CBL-CSBL-04 CBL-CSRD-04	4	370	0.07	0.23	0.40	150	0.15
CBL-CSBL-02 CBL-CSRD-02	2	634	0.07	0.30	0.46	205	0.24

Windynation welding cable

We need to double this setup for F1 and F2 because we have two charge controllers.

Let's calculate the wire from the busbar to the inverter/charger. The inverter's maximum power draw is 5kW.

$$\frac{5,000W}{48V} = 104A \times 1.25 = 130A$$

The closest fuse to 130A is a 150A MEGA fuse (F3). A cable that can carry at least 150A is 2AWG or 35mm². Again, the MEGA fuse needs to be rated at 70VDC.

Let's calculate the wire and fuse size from the busbar to the batteries. It will be higher than the current from the battery to the inverter. The charge controller can deliver 100A each.

$$100A \times 2charge\ controllers \times 1.25 = 250A$$

Since this will be a large cable, we can split it in two. Each cable should be able to carry 125A. I recommend adding a little bit more safety margin to account for imbalances between the two cables. You should also make these the same length.

Let's fuse these cables with two 150A Class-T fuses in parallel. A cable that can carry at least 150A is a 2AWG or 35mm² welding cable, which is rated to carry 205A.

windynation

PowerFlex Welding & Battery Cable Specifications

Part # BL = Black RD = Red	Size (AWG)	Stranding (0.25mm)	Insulation Thickness (inches)	Conductor Diameter (inches)	Approx. Total Diameter (inches)	Maximum Amperage	Approximate Weight (lbs/ft)
CBL-CSBL-08 CBL-CSRD-08	8	131	0.07	0.129	0.31	55	0.07
CBL-CSBL-06 CBL-CSRD-06	6	263	0.07	0.20	0.36	115	0.11
CBL-CSBL-04 CBL-CSRD-04	4	370	0.07	0.23	0.40	150	0.15
CBL-CSBL-02 CBL-CSRD-02	2	634	0.07	0.30	0.46	205	0.24
CBL-CSBL-10 CBL-CSRD-10	1/0	1004	0.08	0.37	0.56	285	0.38

Selecting a welding cable from Windynation

The AC loads should be wired like the previous system.

The ground neutral bond is handled automatically by the Multiplus 2.

I recommend using the Victron MEGA fuse holder with a busbar.

Recommended Brands

The following brands are my recommendations. These companies have a good reputation within the solar community. The brands are listed in no particular order.

Epever

Epever is a Chinese company established in Shenzhen that manufactures charge controllers and off-grid inverters. Available charge controllers can either be MPPT or PWM based, and their outputs can go from 10A to 60A. The brand also makes pure sine wave inverters.

Rich Solar

Rich Solar is a popular American-based company dedicated to supplying off-grid solar equipment for RVs, agriculture, housing, marine, light, and heavy industrial equipment. The brand manufactures solar panels (both rigid and flexible), charge controllers (both MPPT and PWM), inverters, solar lights, and accessories and sells pre-assembled solar power kits.

Victron Energy

Victron Energy is a company from the Netherlands that is one of the top brands available in the market for off-grid, grid-tied, and commercial systems. I recommend their charge controller, DC to DC battery charger and inverter/chargers (multiplus and quattro). Their batteries are quite expensive compared to the rest of the battery market.

Renogy

Probably the most popular option for RVs in the off-grid market is Renogy. An American-based company that manufactures charge controllers, pure sine wave inverters, deep cycle, lithium-ion phosphate batteries, and solar panels (flexible and rigid). Be careful with their lithium batteries. They cannot be put into series.

AltE Store

Established in 1999, altE Store is a Massachusetts-based retailer specializing in renewable energy products. They offer a wide range of solar panels, inverters, batteries, and related components, catering to both DIY enthusiasts and professional installers. Their mission is to make renewable energy accessible and affordable, providing educational resources and technical support to customers.

Battleborn

Founded in 2013 and based in Reno, Nevada, Battle Born (a Dragonfly Energy brand) offers durable LiFePO$_4$ batteries for RV, marine, and off-grid systems. Known for reliability and customer support, they provide a 10-year warranty and focus on sustainable energy solutions. However, they are quite expensive compared to similar batteries. If you want service, use battleborn.

Growatt

Established in 2011, Growatt is a top provider of distributed energy solutions known for PV inverters, energy storage systems, EV chargers, and smart energy management. With a global presence in over 150 countries, Growatt ranks among the top three global PV inverter suppliers and is the leading provider of residential inverters worldwide.

Litime, Redodo, and Powerqueen

LiTime: Formerly Ampere Time, rebranded in 2022, LiTime specializes in lithium iron phosphate (LiFePO$_4$) batteries for RVs, marine, solar, and off-grid systems. Known for lightweight, durable batteries, the brand generally has positive reviews.

Redodo: Focused on LiFePO$_4$ batteries, Redodo caters to solar storage, RV, and off-grid markets. Their compact, reliable batteries are praised for durability and efficient power delivery, making them popular for energy storage.

Power Queen is a company specializing in lithium iron phosphate (LiFePO$_4$) batteries, offering reliable and efficient energy storage solutions for various applications.

Giandel

Established in 2012, Giandel is a prominent manufacturer specializing in power inverters, particularly pure sine wave and modified sine wave models. The company is committed to integrating innovation and quality into its products, offering a range of inverters with varying power capacities to meet diverse energy needs.

EG4

EG4 Electronics specializes in advanced solar energy components, including batteries, inverters, and energy storage systems. Their products are designed to enhance energy independence and efficiency for residential and commercial applications. EG4 is committed to delivering high-quality, cost-effective solutions that empower users to manage their energy needs effectively. The EG4 6000XP, 12kPV, and 18kPV hybrid inverters are popular products.

Current Connected

Based in Las Vegas, Nevada, Current Connected is a family-owned company dedicated to promoting energy independence through renewable energy solutions. They offer a range of products, including solar panels, inverters, and batteries, catering to both DIY enthusiasts and professionals. They sell done for your cables which are cut at your desired length and you can choose the lugs on either end as well. So, if you don't want to crimp cables yourself, this is the go-to solution.

SanTan Solar

SanTan Solar, established in 2015 and headquartered in Gilbert, Arizona, specializes in providing new, used, refurbished, and overstock solar panels. The company offers consultation services, assisting customers in determining their solar panel needs, and handles selection and transportation of products. SanTan Solar is committed to making solar energy accessible to a broad audience by offering affordable options for various budgets.

Docan Energy, Shenzhen Qishou, Exliporc New Energy

When sourcing raw 3.2V $LiFePO_4$ cells, Docan Technology (Shenzhen) Co., Ltd., Shenzhen Qishou Technology Co., Ltd., and Exliporc New Energy (Shenzhen) Co., Ltd. are all suppliers that I have used in the past and that I recommend other people use. It's best to contact them through Alibaba.

Windynation

This is a store on amazon which sells welding cables. I recommend them for all your DC cabling (they also have PV cable). They sell lugs as well. Do not buy cheap lugs, use lugs that are at least $1 each.

Conclusion

This book's goal is to guide you through the essentials of setting up your own off-grid solar power system.

I hope it has served that purpose and given you the confidence to bring your solar project to life.

Remember, safety always comes first. Take great care when handling electrical tools and equipment and ensure you use insulated tools whenever possible.

Embrace the journey of building your solar system from scratch. If you're uncertain about any part of the process, don't hesitate to consult a licensed electrician.

You're also welcome to reach out to me directly at cleversolarpower.com

You will find more diagrams there as well.

For full-color images and schematics from this book, visit cleversolarpower.com/offgridsolarbook
password: **offgrid**

Best of luck on your journey to energy independence!
Take care,
Nick

Made in the USA
Las Vegas, NV
06 December 2024

12818105R00213